T0176720

Protection of Substation Critical Equipment Against Intentional Electromagnetic Threats

Protection of Substation Critical Equipment Against Intentional Electromagnetic Threats

Vladimir Gurevich

Translated by
Kevin Bridge BA MCIL

Authorized Translation from the Russian language edition published by Infra-Engineering

First published under the title *Protection of Power Substation Equipment against Electromagnetic Pulse* by Infra-Engineering, Moscow.

This edition first published 2017

Registered Office
John Wiley & Sons Ltd, The Atrium, Southern Gate, Chichester, West Sussex, PO19 8SQ, United Kingdom

For details of our global editorial offices, for customer services and for information about how to apply for permission to reuse the copyright material in this book please see our website at www.wiley.com.

Library of Congress Cataloging-in-Publication data applied for

ISBN: 9781119271437

A catalogue record for this book is available from the British Library.

Cover image: Powerplant: Kittikorn/Gettyimages
Universe: PlanilAstro/Shutterstock

Set in 10/12pt Warnock by SPi Global, Pondicherry, India
Printed and bound in Malaysia by Vivar Printing Sdn Bhd

10 9 8 7 6 5 4 3 2 1

Contents

About the Author

Vladimir Gurevich was born in the city of Kharkov (in the Ukraine) in 1956. In 1978 he graduated from the Faculty of Electrification at the P. Vasilenko Kharkov National Technical University. From 1980 to 1983 he was a post-graduate student. In 1986 he defended his PhD thesis: 'Quasistatic Switching and Regulating Equipment with High-Voltage Decoupling' at the National Technical University, Kharkov Polytechnical Institute.

He worked as a Lecturer and acting Associate Professor at the P. Vasilenko Kharkov National Technical University and as Chief Engineer and Director of the Scientific Technical Enterprise 'Inventor' (in the city of Kharkov). He led a number of projects to develop new types of equipment, which were conducted at the behest of the Defence Sector Industries in the USSR. Following the collapse of the USSR he has been engaged in developing and administering the production of automation equipment for electrical power systems. He now works for the Israel Electric Corporation as a Senior Specialist, and Head of section at the Central Electric Laboratory. Since 2006 he has been Professor Emeritus at the P. Vasilenko Kharkov National Technical University. Since 2007 he has been an expert on the TC-94 Committee of the International Electrotechnical Commission (IEC).

V.I. Gurevich has written 13 books (six of which have been published in the USA), more than 120 patents and more then 200 scientific and technical articles.

Preface

Some 20 years ago, any mention of the electromagnetic pulse (EMP) from a nuclear explosion could only be found in civil defence pamphlets. Moreover, this was just a brief mention and no more. Therefore, this pulse was perceived as something very exotic and was little understood. The military were, naturally, very well aware of this effect of a nuclear explosion, but all information on this topic was meticulously classified. At that time this was completely justified considering the technical difficulties and material costs involved in obtaining this information. However, as a result of this policy civilian specialists working in the various technical sectors had no idea until recently about this phenomenon, or of the dangers that it posed (and some are still not aware even now).

In the meantime, the modern trend in the development of technology involving an expansion in the universal application of microelectronics, microprocessors, computers and the rapid growth in the productivity of microprocessors accompanied by a dramatic increase in the number of micro-transistors per unit volume and a reduction in the operational voltage and in the levels of insulation between the internal elements and layers in silicon chips has, on the one hand, led to a dramatic increase in the vulnerability of modern technology to EMP and, on the other hand, to a stimulation of interest on the part of the military in using EMP as a self-sufficient and highly effective weapon. If in the past this surprising aspect of a nuclear explosion was only of interest to the military in the context of targeting the electrical systems in an enemy's aircraft and rockets accurately using air defence forces (the warheads found in many of the rockets in the different Air Defence systems, even the short range ones, are fitted with nuclear charges), then contemporary thinking is that EMP represents an ideal non-lethal weapon that is capable of taking almost the entire infrastructure of an enemy out of action by detonating a nuclear charge at high altitude without killing a large number of people. This inspired the military to such an extent that they commissioned the development of a nuclear charge with an enhanced electromagnetic pulse effect; the so-called 'super-EMP'. Parallel to this development work began at an accelerated pace on a purely electromagnetic weapon, in which powerful electromagnetic radiation is generated by non-nuclear means and is used to target modern microelectronic and microprocessor systems. Electromagnetic bombs, shells, grenades and rockets with electromagnetic warheads, mobile installations fitted with a wheeled or tracked chassis, providing powerful concentrated radiation capable of striking electronics at long range: all this has long since ceased to be fantasy and is a reality of our times. Regretfully, it can be said that these realities still fail to attract the attention of specialists in many technical fields, particularly in electrical power engineering. This, however, is the basis of a country's infrastructure without which the water supply and communications systems or any other essential services would not be able to function.

A series of previous articles and books written by the author have drawn the attention of specialists to the importance of this problem in connection with the growing threat of the destruction of the electrical power engineering system by these weapons. In this book the emphasis is on practical recommendations to protect the electrical equipment in substations from intentional destructive electromagnetic threats, including a High-altitude Electromagnetic Pulse (HEMP).

It is worth noting specifically that the protection of substations (and indeed other electrical power engineering assets) from these threats is an issue that is not only of concern to those working in energy but also to the microelectronic and microprocessor manufacturing industry producing equipment for power engineering. Therefore, the recommendations set out in this book are intended not only for personnel engaged in operating this electrical equipment but also for those producing it. First and foremost, this is intended for the DPR manufacturing industry, specialists working in engineering companies, the heads of the electrical power engineering sector, and also lecturers, and postgraduate and undergraduate students specializing in electrical power engineering subjects at universities.

Only through the combined efforts of specialists is it possible to counter this upcoming threat.

Please forward feedback on the book to the author using the following email address: vladimir.gurevich@gmail.com

1

Technical Progress and Its Consequences

The Philosophy Behind Technical Progress

Rational planning in the development of technology more often than not leads to irrational consequences, and technology enters the human consciousness not as a neutral means of meeting our own needs, but as a goal in itself, an alienated force.
Professor N.V. Popkova, DSc

What is technical progress? The dictionary of philosophy provides this definition:

Technical progress – is the interdependent, and mutually stimulating development of science and technology. This concept was introduced in the 20th Century in the context of a basis that made use of a consumerist attitude to nature and a traditional scientific and engineering view of the world. The aim of technical progress is defined as meeting man's ever growing needs; the means by which these demands are met lies in the realisation of achievements in the natural sciences and in technology.

As N.V. Popkova, Doctor of science in Philosophy wrote in his article 'The Philosophy of Technology' [1.1], technological innovation was indeed introduced by man as a way of improving our daily lives and of meeting our needs: the anthropogenic environment performs this task and enables Earth's ever growing population to obtain the material pre-requisites for life. In recent years, however, ever more profound consequences of technological growth have come to light: the suppression of the inherent biological and humanitarian aspects of human life, and their displacement with anthropogenic values and arrogance. This gives rise to an ambiguous evaluation of the role of the anthropogenic environment: the predominantly positive evaluation that existed in the past and the negative one, which is gaining weight. The main problem lies in the intricacies of managing the anthropogenic environment and in the fact that it is impossible to control its development or even predict how it will react to the introduction of subsequent innovations. The discovery at every stage of technical work of unpredictable and undesirable results shows that: *the anthropogenic environment has always in part been outside the control of the human race that is creating it, which means that it has always possessed autonomy.*

Protection of Substation Critical Equipment Against Intentional Electromagnetic Threats,
First Edition. Vladimir Gurevich.

Thus it is far from the case that the development of technology has always been aimed at 'meeting the ever growing needs of man', since according to our observations technical progress only began to adopt this characteristic in the second half of the twentieth century.

An old science fiction novel featured an engaging plot, which arose out of something relatively innocent: an unusual night time phone call made to each of the inhabitants of planet Earth. It was in this phone call that the 'Global Mind' announced its' coming to everyone on planet Earth. It turned out that at some stage in its development the proliferation of computers had transformed into something new: millions of computers, which had been combined into an overall network and which controlled everyone and everything on planet Earth had suddenly come to the realization that they represented a single entity capable of reproducing themselves using automated factories and robots that had been integrated into this same network, and of defence with the help of computerized weapons systems designed to destroy mankind. As far as the 'Global Mind' was concerned humanity was nothing more than a rudiment, or ballast that was devouring the planet's resources. You can work out how the plot unfolded from there for yourselves.

Today, almost all modern industrial production methods as well as systems controlling the supply of water, electricity and telecommunications and communications systems, are controlled by computers with a network connection. The terms *Smart Grid* and *Artificial Intelligence based relay protection* have appeared in technical rather than science fiction literature. Issues surrounding the creation of a Smart House, in which even the fridge would be able to assess the levels of the provisions stored inside it and on the basis of this analysis of demand draw up an order and send it via the network to the local supermarket, are being discussed today in technical literature and not in science fiction. Today microprocessors can be found anywhere, even in the toilet seat lid.

Humanity is making huge strides towards the creation of an unpredictable Global Mind, which the old science fiction novel had foreseen. Thus this old plot has long since made the leap from the pages of science fiction novels into the pages of respected philosophical journals and books that illuminate issues in the philosophy of technology. This is a relatively new field of philosophical research, which is aimed at understanding the nature of technology and evaluating its impact on society, culture and man. One school of thought suggests that the philosophy of technology is not, if anything a philosophy in itself but a multidisciplinary intellectual field, in which technology as well as the problems it creates are typically examined as broadly as possible.

At the VISION-21 symposium that was conducted in 1993 by NASA's Lewis Research Centre and the Ohio Aerospace Institute the famous professor of mathematics Vernor Vinge delivered a much talked about speech [1.2]:

> *The acceleration of technical progress – is the key feature of the XX Century. We are on the verge of changes comparable to the emergence of man on Earth. The specific reason for these changes lies in the fact that the development of technology inevitably leads to the creation of beings with an intellect that surpasses that of humans...Large computer networks (and their consolidated users) are able to 'come to the realisation' that they are supernaturally intelligent beings... an event like this would nullify the entire statute book of human laws, possibly in the blink of an eye. An uncontrolled chain reaction would begin to develop exponentially with no hope of regaining control of the situation.*

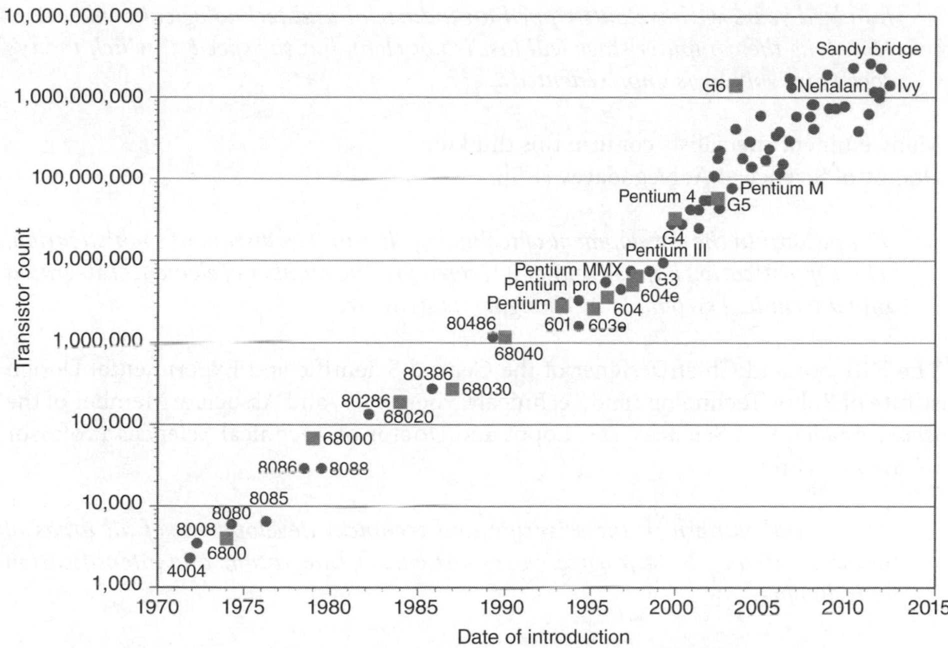

Fig. 1.1 The relationship between time and the number of transistors in microprocessor chips. The vertical axis has a logarithmic scale and the relationship conforms to exponential law.

Vinge proposed a new term for this phenomenon: *Technological singularity*. Normally singularity is understood to mean an isolated point of some kind or a function field, the meaning of which denotes infinity or which demonstrates other behavioural irregularities, it denotes a critical point beyond which the value of a function becomes indefinite and unpredictable. Typical examples of singularity are an avalanche breakdown in semiconductor structures, a tunnelling effect in electrical contacts and in semiconductors, an area of volt-ampere response in a negative resistance diode and so on. Technological singularity implies a certain point in the development of technology as a whole, but specifically the development of computer technology and artificial intelligence beyond which their further development becomes firstly irreversible and independent of humans, and secondly unpredictable.

Naturally, the so-called Moore's law [1.3] would have influenced Vinge's views; this was formulated in 1965 by one of the founders of Intel Gordon Moore. This law states that the number of transistors in microprocessors doubles approximately every 2 years and their productivity grows exponentially as in Fig. 1.1. This law has been valid for 40 years now. Not only do microprocessor and computer technology, which are becoming ever more complex, conform to exponential law but also other types of technology, and with it society. The sociologist M. Sukharev in his work 'An Explosion of Complexity' [1.4] writes:

> *There is another pattern that is visible in the development of society - the acceleration in the growth of complexity over time. Tribal people have lived on the Earth for thousands of years, armed with spears and arrows. In the space of a few*

hundred years we have outstripped an industrial and technological civilisation. How long the computer stage will last is not clear, but the speed at which today's society is evolving is unprecedented…

Many eminent specialists confirm this thinking:
Doctor of Sciences I.A. Negodayev [1.5]:

The pattern in the development of technology lies in its subsequent sophistication. This sophistication happens either by increasing the number of elements integrated into a technical system, or by changing its structure.

The Director and Chief Designer of the Central Scientific and Experimental Design Institute of Robot Technology and Technical Cybernetics, and Associate Member of the Russian Academy of Sciences V.A. Lopot and Doctor of Technical Sciences Professor E.I. Yurevich [1.6]:

The overall pattern in the scientific and technical development of all areas of human activity – is the progressive sophistication, integration, and intensification of technology.

Bezmenov A.E. PhD [1.7]:

The trend in the development of technology is characterised by the ever growing sophistication of machines, equipment, and installations. With an increase in the sophistication of these items, their reliability (all other things being equal) diminishes.

If the 'Explosion of Complexity' in everyday technology is happening to everyone in plain sight and requires no evidence, then the sophistication of technology in industry is not so obvious to the layman. Therefore, we will examine a few concrete examples that confirm this trend.

The Swedish company Programma Electric AB, known all over the world, was founded in 1976 (this company was acquired by General Electric in 2001, and in 2007 it became part of the Megger Group Ltd) and produces a huge nomenclature of equipment and installations to test electrical power engineering equipment: from highly accurate timers and systems testers of protective relays to sources of powerful currents. One of the items this company produces is the B10E equipment pictured in Fig. 1.2, used to measure the minimal pick up voltage in high voltage circuit breaker drives.

In accordance with IEC standard 62271-100 these circuit breakers need to be tested for their compliance with the manufacturer's parameters for the minimal pick up voltage. In general, this refers to a swash that performs a very simple function: a preliminary check on a certain level of voltage controlled by a voltmeter, with the voltage being fed subsequently to the device's output terminals. It is not complicated to develop a diagram for this device, as in Fig. 1.3. In this device the output voltage is set by the variac AT, rectified by a diode bridge, and smoothed by the large capacity (several thousand microfarads) capacitor C. The voltage is fed to one pair of output terminals from a variable alternating current source and to the other from a variable direct voltage source.

Fig. 1.2 The outside of a B10E type device used to test the minimal pick up voltage in high voltage circuit breaker drives.

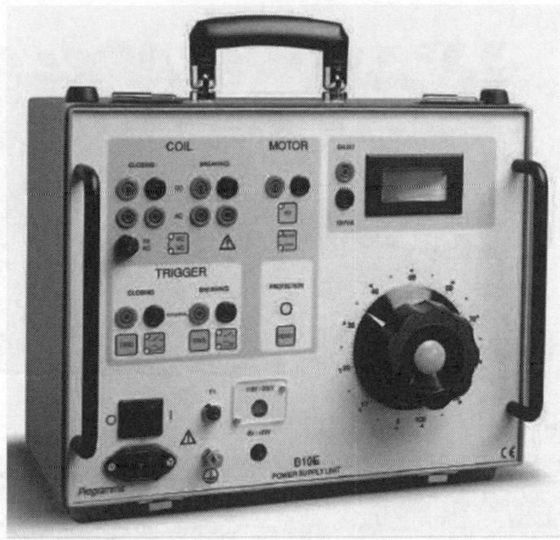

Fig. 1.3 An example of a diagram for a simple device used to test high voltage circuit breakers, which performs all the necessary functions.

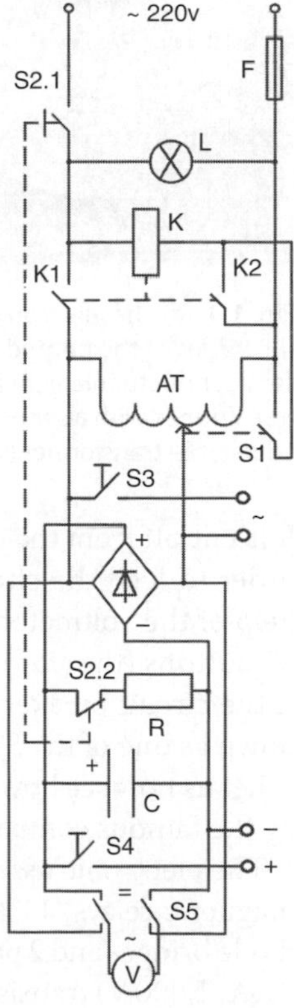

The output voltages are monitored with the help of the voltmeter V. In order to prevent any inadvertent high voltage (250 V) feed from the device to the low voltage (24–48 V) coil or to the motor, the S1 micro-switch is fitted to the variac in such a way that its contacts are closed by the movement of the plunger attached to the shaft and only in the neutral position by the variac arm. When the S2 button is pressed the discharge resistor

(a)

(b)

Fig. 1.4 (a) The electronic assembly of the B10E device. The semiconductor elements installed along the edges of the printed board are pressed against the case, which is used as the heatsink for the semiconductor elements during assembly. (b) The power unit in the B10E device: 1 – the multi circuit transformer with a series of different output voltages to feed the device's electronic assemblies; 2 – an adjustable transformer (variac) and 3 – the sensor board for the angle sensor on the variac shaft.

R is cut off from the capacitor C and the voltage is fed to the device's input terminal. In order to feed the circuit breaker coil with the voltage preliminarily supplied with the help of the voltmeter and variac, in addition to the S2 button being pressed, one of the S3 buttons (the alternating current output) or S4 (the direct current output) is pressed. If the circuit breaker does not work, the voltage is increased and the S2 button is held down as one of the S3 or the S4 buttons is pressed once again.

Let us now see how this simplest of algorithms is realized in the B10E device produced by the famous company in Fig. 1.4.

The electronic assembly of the B10E device shown in Fig. 1.4(a) contains 13 electromagnetic relays, 14 different types of integral microcircuits, 10 1A current rectifying diode bridges and 2 powerful 40EPS08 (40A, 800 V) type diodes, 4 high power BUX98AP (24A, 1,000 V) transistors); 3 high power BTA26–400B (25A, 400 V) triacs, 4 high power gate-turn off GTO thyristors (13.5A, 800 V) and 2 precision PBV type current shunts.

To be honest, I admit that when I opened this device with the aim of repairing it, I was completely shocked by what I saw. I was particularly affected by the electronic angle sensor on the variac shaft in place of the simplest micro-switch (as shown in Fig. 1.3). The complete incompatibility of the simplest of functions carried out by this equipment with its technical realization is plain to see. It would be interesting to know what justification the developers of this device used for such an agglomeration of electronics.

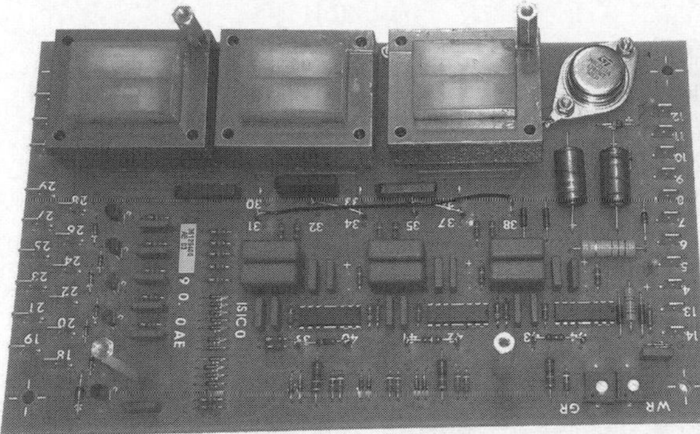

Fig. 1.5 The two modules of the old charger controller that were developed and mass produced in huge numbers in the 1970s by AEG. On the right is the analogue module that controlled the output voltage and current of the charger and which sent a signal to the pulse firing module (on the left) and formed control pulses for the thyristors.

Fig. 1.6 A battery charger's control module produced on a modern element base in accordance with a design developed by AEG in the 1970s.

Here is another example from the field of power stationary battery chargers, widely used in power stations and substations in auxiliary direct current systems. This unit consists of the following principle assemblies: a power transformer, a block of power thyristors and an electronic thyristor control assembly. At the beginning of the 1970s AEG developed a thyristor controlled charger, shown in Fig. 1.5, which proved so successful that it is still used more than 40 years later by different manufacturers in various types of charger unit. Moreover, some manufacturers have copied this control assembly in its entirety, while others have transferred it over to a modern element base, see Fig. 1.6, which in essence does not change the assembly.

Unfortunately, no matter how well analogue technology has proved itself in charger control systems over the course of 45 years both in terms of reliability and ease of repair, at this point it has to be said that it has already been usurped completely by digital devices

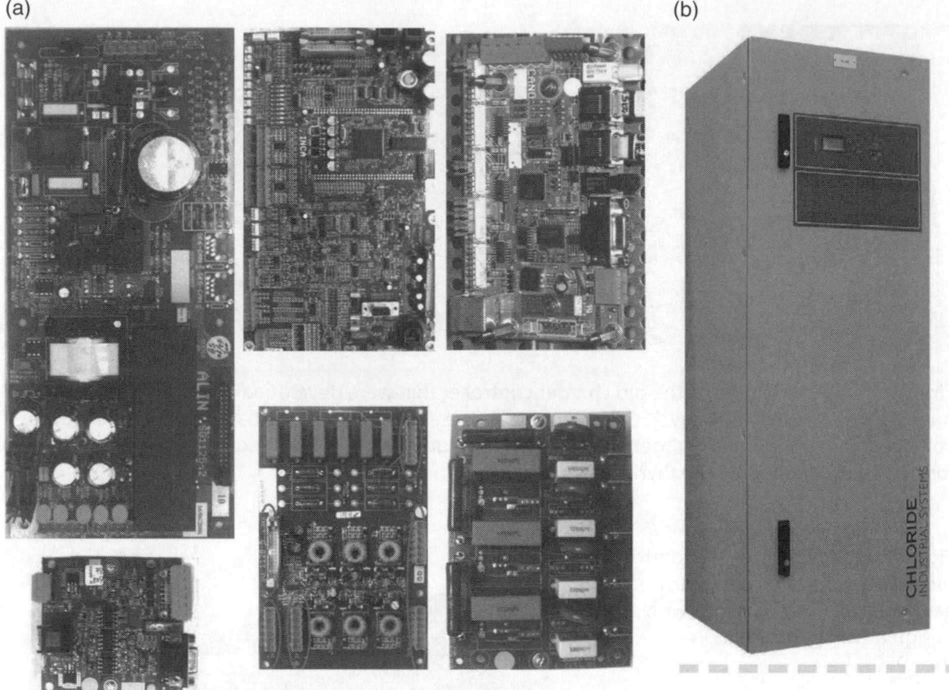

Fig. 1.7 (a) A set of the microprocessor based control modules for the Apodys series battery charger produced by Chloride France S.A. (b) A battery charger from the Apodys series produced by Chloride France S.A., part of the Emerson group.

based on microprocessors. What in terms of new properties have microprocessor controlled battery chargers acquired? (See Fig. 1.7.)

They are as follows: a 'branching' menu in which it is not easy to find the required function in place of the three output voltage control potentiometers and a mode switch; an IP-address and a network connection, which enables hackers to interfere with the operation of the unit; a modern fiber optic connection between the internal modules in place of the traditional copper cables and so on.

One might continue the description of examples that demonstrate 'an explosion of complexity'. For example, in Fig. 1.8 the MCT1600 device, produced by Megger and designed to measure insulating resistance, the transformer ratio and the knee point of the volt-ampere characteristics for the current transformer, which when switched on boots up a full scale VX Works operating system (a 64-byte real time operating system), is a case in point.

As are the insulation resistance meters produced by the same firm, which underwent evolution from a miniature device with a generator that was turned by a handle into extremely complex microprocessor based assemblies, see Fig. 1.9.

A typical example of the 'explosion of complexity' in electrical power engineering is the Smart Grid. It is well known that the Smart Grid concept presupposes the installation of microprocessors in all the elements without exception of the electrical engineering production, distribution and metering system as well as the establishment

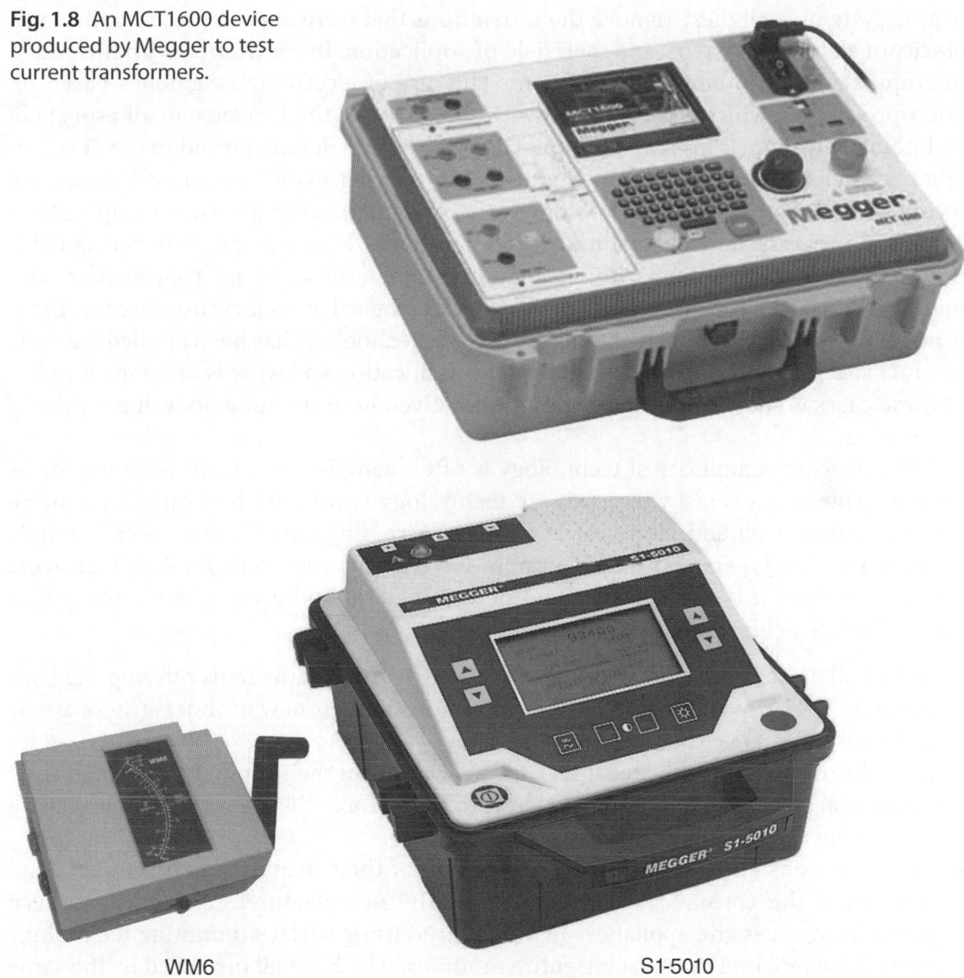

Fig. 1.8 An MCT1600 device produced by Megger to test current transformers.

Fig. 1.9 Insulation resistance metering equipment produced by Megger: WM6 is the simplest device fitted with a generator and a handle with which to rotate the armature; while the S1-5010 is the most complex microprocessor based device.

of information channels between these elements based on computer networks, predominantly Wi-Fi. The idea of the proponents of Smart Grid was that the energy system of the future should resemble a modern, sophisticated network computer game with thousands of component participants playing a role in the electricity networks. One of the central participants in this 'game' are DPRs with an artificial intelligence and which are self-adaptable, with an indeterministic logic that looks ahead, that is to say it acts independently and at its discretion [1.8].

The affordability and accessibility of microprocessors, industrial control equipment, modern highly integrated electronic components, as well as the huge and ever expanding nomenclature of these components on the market together with the exceptionally high productivity of this equipment, designed for automatic installation and soldering of surface mounting components onto a circuit board and automatic circuit board testing systems – all these remove the restrictions that were once in place on the complexity of electronic systems and their field of application. In connection with this today, microprocessors can be found anywhere. This use of electronic assemblies based on microprocessors, which has expanded with the speed of the Universe in all aspects of technology given their insatiable complexity, is today the defining trend in the development of technology. The proponents of technical progress as we know it today are trying to convince us that technology's unceasing and ever growing complexity is 'technical progress' in itself. Naturally there are some technical and engineering fields that are unable to function without computer operations and microprocessors and microprocessor technology really has made technological leaps forward possible. In far from all cases, however, in which microprocessor technology has been applied have the product specifications provided a reasonable justification and what is more the number of these cases is snowballing and the examples given here are but a poor illustration of this process.

If the growing complexity of technology is often completely unjustified, however, as we have shown previously, then why is technology constantly becoming ever more complicated of itself and moreover at an ever increasing pace? The answer is simple enough: the developers and manufacturers have an interest in technology's growing complexity since it is this constant and determined complication of technology that allows them to achieve certain goals all at once:

- First of all, to raise the effectiveness of their advertising campaigns offering the consumer an ever growing number of new functions in their new products (these are by no means always necessary);
- Secondly to undermine the reliability and service life of their products (which in itself is a natural result of complication), that is to say to force the consumer to purchase a new product more often;
- Thirdly to constantly reduce the serviceability of their manufactured products and to increase the consumer's dependence on the manufacturer. The most modern electronic devices and appliances manufactured using surface mounting technology can only be repaired by replacing entire modules, which are all produced by the same manufacturer.

In many cases the cost of these modules is disproportionately high even though the consumer is forced to purchase them at a clearly inflated price. Thus in many cases the complication of technology has become an artificial process, which often does not change the effective basis initiated by the manufacturer with the aim of further enrichment.

How shameless, however, is this process of technological development?

In the words of Doctor of Technical Sciences and Professor, and head of Central Scientific Research Centre 46 of the Department of Defence of the Russian Federation Major-General V.M. Burenok: a distinguished figure in Russian science [1.9].

> *Technological development conceals within it a multitude of threats, which in terms of their variety and the repercussions of their influence are unpredictable for the fate of civilisation...In the last few years scientific and technological progress has brought the world many technical benefits, but with them a persistent headache. Examples are: computer technology and cyber terrorism, modern*

information communication systems and information wars, complex infrastructure and technical asset management systems as well as the serious consequences when they fail, a knowledge of the basis of life as well as genetically modified products, and the advent of the potential for the artificial cultivation of dangerous viruses and so on. Moreover many of these threats that have been generated by new opportunities in technology have not come about straight away and could not have been predicted (either that or those that forecasted them were labelled irrelevant fantasists or eccentrics, that were not worth taking seriously).

This, however, is what the Academician N.N. Moiseyev wrote on the subject:

...scientific and technical progress, and the growth in the power of civilisation does not just bring benefits. The power that this gives people also has to be used wisely. Today man finds himself in Gulliver's position, when he visited the Lilliputian crystal shop. One false move and this whole crystal wonder would turn into a heap of broken glass.

Knowing that these dangers exist it would most likely be possible to try and prevent them. This, however, is what the distinguished specialist already cited previously had to say on the subject [1.9].

Even when the layout of a technical system of some kind has long since been drawn up, but new threats have emerged it seems it is no easy task to predict this situation. Rarely does an analyst for example undertake to predict the consequences of a cyber attack on for example a nuclear power station or a large hydroelectric station's control system or on an air or railway traffic control system. Forecasts such as 'this is going to be awful,' and 'huge and unavoidable losses,' don't suit anyone but assessments such as 'the likelihood of the release of an amount N of radioactive material into the atmosphere will rise to...,' or 'the number of aircraft accidents in airspace with a probability of p will reach the value K' are very hard to make. In order to do this (to make a forecast) a model of the system (or asset) is required, which is almost comparable to the real system, together with a knowledge of how far a hacker's skill has developed, as well as a means of penetrating the system under attack and so on. Firstly, however, this is almost impossible to do, and secondly if this model existed and were to fall into the hands of intruders (hackers) then this makes the chances of the system operating trouble free highly elusive.

He is supported by the well-known astrophysicist L.M. Gindilis, who wrote in his work [1.10]:

The acuteness of the situation lies in the fact that the collapse should come very quickly, in the first few decades of the XXI Century. Therefore, even if humanity were aware of how to 'avert' (or even to stop) this process and had the means and the motive even to turn a corner today – there would still be insufficient time remaining, since all the negative processes possess a certain momentum, which means they cannot be stopped immediately...the global economy is like a heavily

> *loaded vehicle moving at high speed towards the end of the road, straight to the abyss. Evidently we have already passed the point at which we should have turned to enter a 'turning trajectory'. We won't have time to slow down either. The situation is exacerbated by the fact that nobody knows where the wheel or the brake is. Nevertheless, both the crew and the passengers are very complacent, naively supposing that 'when necessary' they will be able to figure this vehicle out and perform the required manoeuvre.*

In conclusion, we cite the words of the proponent of the theory of technological singularity Vernor Vinge:

> *If 'Technological Singularity' is destined to be, then it will happen. Even if all the nations acknowledge the 'threat' and are scared to death – the progress will not stop. A competitive advantage – be it economic, military, or even in the arts – or any achievement in automation systems would be so overwhelming that a ban on similar technology would merely guarantee that someone else would get there first. I have already expressed my doubt that we will succeed in preventing Singularity, and that its coming is an inevitable consequence of man's natural competitiveness and of the potential inherent in technology.*

The natural (if this term can be used in a technical context) development of technology and engineering was examined previously. There is, though, another side to this problem, which has never been considered in the philosophy of technology. This concerns weapons capable of destroying technology that have developed alongside it. As technology has become more complicated and ever more 'electronic' and 'computerized', so has its vulnerability to intentional remote destructive threats, including cybernetic and electromagnetic threats [1.11]. Therefore, weapons system developers are paying more and more attention to creating new types of weapons aimed at striking technology exclusively, as opposed to humans. This is also a component of 'technical progress', which has been undeservedly excluded from consideration under the philosophy of technology. After all, the sudden destruction of complex electrical systems and the computer networks that branch out from them, and on which modern civilization is founded, could lead to the collapse of that same civilization.

Thus for contemporary society there are not one but two opposing and highly dangerous trends: how uncontrolled development leads to singularity, and the ever growing danger of the sudden and deliberate destruction of this same modern technology using special types of weapons.

1.1 Technical Progress in Relay Protection

Over the course of hundreds of years, electromechanical protective relays (EMPR) have provided solutions to challenges that have arisen in relay protection and, bearing in mind that they comprise around 70–80% of all the protection systems used around the world, then it can be said with some certainty that electromechanical relays today are capable of solving all the challenges facing relay protection. Nevertheless, the last 20–30 years have seen electromechanical relays being replaced universally by Digital

Protective Relays (DPR) and numerous programmable logic controllers (PLC), which control the modes of operation of electrical equipment, have resolutely entered our lives and in many cases it is not possible for electrical power engineering to function properly without them. This is not about certain unique capabilities inherent in microprocessor technology but about the established trend, conditioned by a variety of reasons including bumper profits, obtained through the fully automated production of printed circuit boards (PCB) of DPRs, compared to the production of the previous generation of precision mechanical type relay protection. The search for ways to reduce the production costs and to increase the profitability of production have led to the development of new types of EMPR ceasing some 30–40 years ago and all the efforts of the developers being directed towards the creation to begin with of semiconductor static, and subsequently even microprocessor based digital protective relays (DPR). The first DPR simply copied all the functions and the characteristics of previous generations of relays. New characteristics and scope for DPR only appeared many years later. Therefore, it is doubtful that it can be said that the advent of DPR was conditioned by the actual demand for relay protection. As a result of this technical policy practised by the manufacturers, almost all the world's leading relay protection manufactures completely ceased production of all other types of protection, apart from DPR, and there remained almost no alternative to DPR (one very small exception in this global trend).

The very first examples of DPR, which simply copied the functions of static transistorized semiconductor type relays, see Fig. 1.10, revealed serious problems with the DPR: they would fail on a regular basis and they could not be repaired owing to the presence of a specialized microprocessor and a read only memory (ROM) with a programme written onto it. As a result, if an RXIDF-2H transistorized relay, or one adapted for other discreet components, was repaired relatively quickly and set to work again then their microprocessor analogue: RXIDK-2H would just have been thrown away. As a result, RXIDF-2H DPRs have long since been withdrawn from production, while the RXIDF-2H relays are still in operation. The trend in the reduction of the reliability of relays in connection with the transfer to DPR, that was observed at the very beginning of this process still continues today, despite the fact that the modern generations of DPR have little in common with the very first examples produced several decades ago, see Fig. 1.10. This is testimony to the fact that the problem lies not in the individual technical shortcomings of the early examples of DPR, but is systemic in nature. However, nobody wanted to be known as a retrograde and nobody wanted to talk about the obvious problems that accompanied the introduction of DPR, which had received nothing but a rapturous response. Furthermore, several billions of dollars have been spent on developing ideas and technology linked to the development of DPR over the last few years across the world and the fact that this line of work has become a highly profitable business for thousands of scientists and engineers, and which has fed them over the course of decades, all the discussions concerning the problems and shortcomings of DPR have either nipped a violent rejection in the bud or have had to face one on the part of representatives of the manufacturing enterprises, scientists, developers, project engineers or anyone else involved in this huge business. An attempt by the author in the past to draw attention to the existence of problems with DPRs gave rise to furious allegations of incompetence, of a lack of understanding of the fundamentals of relay protection and even of trying to delay technical progress. In recent years, it is true a realization of the problems with DPR has come about, but this process is reminiscent of an old joke: to

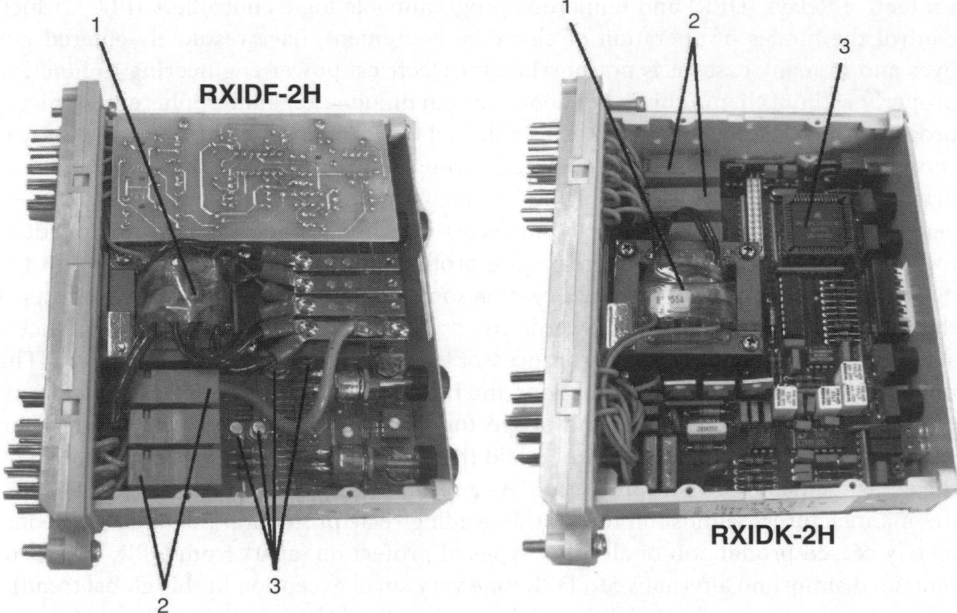

Fig. 1.10 Two dependent time-lag current relays, with identical technical parameters, characteristics, dimensions and manufactured in identical standard COMBIFLEX© casings and produced by the very same company (ABB). On the left is a static semiconductor type RXIDF-2H, while on the right is a microprocessor type RXIDK-2H. 1 – Input current transformer; 2 – electromagnetic output relays and 3 – transistors in a static relay and a specialized microprocessor – in a microprocessor unit.

begin with: 'this can't happen because it would never happen', then 'there is something in this' and, finally, 'could it be any different?' This is without the middle phrase in this process, however, that it to say without an acknowledgement that the first person to shed light on these problems was right.

Many plagiarists have simply copied whole sections of text from the author's books and articles and incorporated them into their own articles without adding any links to the source of the information; they report at conferences and even present this work without any alterations at competitions to find the best student dissertations [1.12–1.19].

1.2 Microprocessors – The Basis of the Contemporary Stage of Technical Progress

The affordability and accessibility of microprocessors, industrial sequence controllers and modern electronic components with a high degree of integration, as well as the enormous and ever growing nomenclature of components such as these available on the market, the exceptionally high productivity of equipment designed for automatic installation and soldering of surface mounting elements onto a printed board and automated printed board testing systems – all this removes the barriers that were previously in place to the complexity of electronic systems and the scope of their use. In connection with this today, microprocessors can be found anywhere, up to and including the toilet seat, where they measure the temperature of a corresponding part of the body and set the water heater for the inbuilt shower so that the water temperature matches the temperature of the respective body part. This use of microprocessor based electronic systems, which is growing at speed, in all fields of technology, together with their ever growing complexity is today the defining trend in the development of technology. It has become acceptable to label this trend 'progress' in the development of technology and engineering. Naturally, there are some fields in technology and engineering that that are unable to function without computer operations and microprocessors and microprocessor technology really has made technological leaps forward possible. However, in far from all cases in which microprocessor technology has been applied have the product specifications provided a reasonable justification and, what is more, the number of these cases is snowballing.

Whilst observing this trend not as an onlooker, but as an insider so to speak, that is to say having been involved in the operation and repair of complex electrical devices designed for industrial applications such as relay protection and powerful battery charging equipment, invertors and convertors, as well as uninterruptable power supply (UPS) and so on, doubts creep in about whether the trend described here really represents technical progress. Why? Because the boom that is being observed today that is conditioned by the acute complication of equipment and the ever growing use of microprocessors in all technical fields is not so much linked to actual demand as much as to an intention by the manufacturers to outperform their competitors at any cost, to make something that nobody has made before and to earn bumper profits. In itself, the desire to create something new or to reduce production costs can only be welcomed if the trend in replacing analogue systems that have proved themselves by working faultlessly for dozens of years in discrete electronic components with microprocessors did not lead to the equipment becoming significantly more complex. Also, if it would mean that this equipment would not become unserviceable, or that its reliability would not be reduced and the cost of maintaining it in working order were not so high and it did not require personnel to be so highly trained. When ordering this equipment all these problems remain in the shadows and they are only encountered when the equipment first becomes operational. This is the cost that consumers are forced to pay for so-called 'progress'; that is to say the reckless and irresponsible complication of technology, which is often conducted without any justification and only serves to appease the fashion for technology and to pursue consumers for bumper profits.

1.3 Smart Grid – A Dangerous Vector of 'Technical Progress' in Power Engineering

Today, there is probably not one branch of the media that has not written some kind of rapturous ode in honour of so-called 'Smart Grid', which is touted as the in-thing in technology fashion, bringing us benefits never seen before. Today only the indifferent keep their counsel about their contribution to the development of this new and fashionable direction. It appears that it is not only microprocessor based electricity meters but arc furnace transformers, reactive power compensating equipment and superconducting electrical cables are even electric, as are all elements of a Smart Grid that require money for their production development. Today, targeted state investment programmes are being drawn up and investments running into billions are being allocated. An enormous mechanism is being set in motion to 'draw off' and disperse funds from state budgets for a line of work for which nobody is even able to provide a full, clear and coherent explanation [1.20]. It is common knowledge, however, that the Smart Grid presupposes the installation of microprocessors in all the elements of the production, distribution and metering of electrical power systems without exception, and the establishment of channels of communication based on computer networks, predominantly Wi-Fi. According to the proponents of Smart Grid the energy system of the future should resemble a modern, sophisticated network computer game with thousands of component participants playing a role in the electricity networks. If the millions of domestic electricity meters that have been incorporated into an overall computer network (that is to say millions of potential points of connection for hackers) are added into this then the entire scale as well as the danger of this undertaking, conditioned by a sharp increase in the vulnerability of the electrical power engineering system to attacks by hackers, computer viruses and intentional, destructive remote electromagnetic threats, which are examined in detail later on, becomes even more obvious. An electromagnetic pulse from a high altitude nuclear explosion conducted in near space above the territory of any particular country is today considered an actual variant of a so-called non-lethal weapon, one capable of taking almost the entire microelectronic infrastructure across the entire territory of a country out of action, but sparing the population their lives.

Alas none of these dangers or simply 'horror stories' as they were disdainfully dubbed by some exponents of 'technical progress' in its contemporary sense, are of much concern to scientists and engineers, who are paid from the Smart Grid development funds. Statements such as this are heard frequently: our task is to further technical progress, and any concern for ensuring the security of the nation's electrical power engineering is the prerogative of the Army and of special forces, so let them take this forward. The inferiority of such an ideology is obvious and does not even require an explanation.

1.4 Dangerous Trends in the Development of Relay Protection Equipment

In a series of previous publications, we have drawn attention more than once to the danger of certain trends in the development of relay protection and dismissed as propaganda by microprocessor based protective relay developers and manufacturers. This refers to the following trends:

1) The unceasing complication of DPR and an increase in the concentration of protective functions in a single terminal [1.21–1.23].
2) Installing functions onto DPR that do not relate to relay protection such as monitoring electrical equipment, for example [1.24, 1.25].
3) Using indeterministic logic in DPR, as well as so-called 'preventive action', which can give rise to the danger of a loss of control over the relay protection functions [1.24, 1.25].
4) The expanding use of freely programmable logic [1.26] in DPR, accompanied by a significant growth in the ratio of mistakes made by operators and of the protection not functioning properly.

5) Making serviceability checks as well as the operation of relay protection more complicated in general in proportion to the rise in the variety of different types of DPR produced by different manufacturers, which differ both in terms of their design and their software, being used in a single energy system. The lack of standards specifying integrated universal requirements for the design and programming of DPR and increasing the intellectual workload on personnel leads to significant economic losses [1.27]. This situation is compounded every year.
6) A significant weakening in the electromagnetic shielding of protective relays and of the energy system as a whole in proportion with an expansion in the use of DPR [1.28–1.30].
7) An increase in the vulnerability of energy systems to attacks by hackers in proportion with an expansion in the use of microprocessor technology and with the use of cheaper networks such as Ethernet and Wi-Fi in place of comparatively well-protected optoelectronic cables in relay protection systems [1.31].

This complication, both in terms of equipment and programming has come at a price. As references [[1.21–1.22, 1.32–1.35] have shown, the transition to DPR has already led to a significant reduction in the reliability of relay protection. Despite this, however, the proponents of DPR think that it there is no need to stop there, but there is a need to continue to make them more complex, increasing the number of functions carried out by a single terminal; using freely-programmable logic in microprocessor based relay protection; and indeterministic logic based on the theory of neural networks; preventive action algorithms; installing information-measuring systems onto DPR and power equipment monitoring systems; using wireless communication channels (Wi-Fi) between the relays and so on. All these new developments, which are financed by large corporations and often even from the state budget, have turned into a vast business and today nobody wants to be excommunicated from this lucrative 'pie'. The players in this business are not concerned in the least about the future consequences of their activity but they are looking to 'push' their new and fashionable ideas onto the market as soon as possible.

Business is business and its framework acts work differently in different countries and in different fields, including in such a sensitive field as relay protection and control in electrical power engineering. You don't believe this? Then familiarize yourself with the motto of the report on the 'Distribution systems of the future: Novel ICT solutions as the backbone for smart distribution' symposium, published in the journal *PAC World*, see Fig. 1.11. The key words here are 'mandatory' and 'urgent'; that is to say without a careful analysis of the long-term consequences of these innovations and without an unnecessary critique. This was how things were done in countries across the world up until very recently.

After a period of very stormy critical reaction to the authors publications and a complete rejection of the negative consequences of the trends set out here in the development of relay protection, in recent years an understanding has come about of the problems that have been set out before by a number of specialists. For example, B. Morris, R. Moxley and C. Kusch (Schweitzer Engineering Laboratories USA) presented the report: 'Then Versus Now: A Comparison of Total Scheme Complexity' at the Second International Conference 'Contemporary Trends in the Development of Relay Protection Systems and the Automation of Energy Systems' (Moscow, 7–10 September 2009) in

by Bernd Michael Buchholz, NTB Technoservice, Germany and Christoph Brunner, it4power, Zug, Switzerland

industry reports

and prosperity of the industry was clearly considered by Madame Merce Griera I Fisa from the European Commission. The SmartGrids are a prerequisite to reach the European 20-20-20 targets in 2020 (*20% improvement of energy efficiency, 20 % share of renewable energy sources to cover the demand of primary energy, 20 % reduction of carbon emissions*). Furthermore, the advanced products and system solutions partly resulting from funded projects will ensure

presented by the 12 participating project teams – beginning with the building automation "SmartHome" and the involvement of household consumers into the electricity market, the automation of distribution networks up to the erection of prospective markets for energy and reserve power. Engaged discussions followed each of the contributions.

The analysis of the consumer behavior in the environment of dynamic tariffs presented a potential

It is mandatory that the new solutions from the project are urgently applied in practice now.

One session considered the barriers for SmartGrid solutions by the current regulation and legal situation in Germany. For many

Fig. 1.11 The motto of one of the publications in a journal that is popular among specialists across the world – *Protection, Automation and Control World* (*PAC World*), September, 2011 (highlighted in a box), reads: 'It is mandatory that the new solutions from the project are urgently applied in practice now.'

which they cast doubt on the need for more and more complexity in protection, arguing their case by using comparative analyses of the reliability of protection based on ever more complicated microprocessor units. V.I. Pulyaev (FGC UES, Russia) also spoke about the poor reliability of DPR at the Third International Conference 'Contemporary Trends in the Development of Relay Protection Systems and the Automation of Energy Systems' (Saint-Petersburg 30 May – 3 June 2011). He noted specifically that a significant proportion of the failures in relay protection occur in microprocessor units (approximately 23% of all cases), which amount overall to around 10% of the total number of protection systems. It goes without saying that this is one of the most important factors that define the need for special measures to be taken to increase the reliability of microprocessor based protection systems. The late Aleksey Shalin (Doctor of Technical Sciences and Professor of the Faculty of Electricity Power Plants at the Novosibirsk State Technical University, and Lead Specialist of the Open Joint Stock Company 'PNP BOLID' in Novosibirsk) wrote openly in his letter responding to one of our publications (see A. Shalin 'Microprocessor Based Relay Protection: The Need for An Analysis of Efficiency and Reliability' in the journal *Electro-technical News*, 2006, No. 2) that the percentage of malfunctions in modern relay protection panels and cabinets often turns out to be significantly higher than for old protection systems based on electromechanical relays, and also that the statistical data confirms the fact of the significant reduction in the effectiveness and reliability during the transition from EMPR to DPR. A. N. Vladimirov (Central Dispatch Administration of Russia's Unified Energy System) also wrote about the reliability problems, as did S. Swain, and D.B. Ghosh (Integrated Electrical Maintenance) among others [1.36].

Stokoe and Gray, in their report 'Development of a Strategy for the Integration of Protection & Control Equipment' at the Seventh International Conference on 'Developments in Power Systems Protection' (Amsterdam, 9–12 April 2001) noted that the old electromechanical relays were durable and long lived systems with a service life of 25 years whereas the service life of modern microprocessor based relays is 15 years or maybe less. They are supported by J. Polimac and A. Rahim (PB Power, United Kingdom) who asserted that the service life of protection systems would decrease from 40 years (for electromechanical systems) to 15–20 years during the transition from electromechanical to microprocessor based relays, and in some cases right down to just a few years following their introduction into service (for microprocessor based systems) [1.36].

The head of the computer division of the Engineering and Technical College at the University of Poona, Maharashtra) Ashok Kumar Tiwari B.E. noted that concentrating a multitude of functions in a single microprocessor terminal reduces the reliability of relay protection sharply, since should this terminal fail a great number of functions would be lost compared to the same scenario if these functions were distributed across several terminals [1.36]. V.A. Yefremov and S.V. Ivanov (of the Engineering Centre 'Bresler') and D.V. Shabanov (FGC UES, Russia) also spoke of the need to limit the number of functions manifested in a single microprocessor based relay protection terminal in their report at the Third International Conference 'Contemporary Trends in the Development of Relay Protection Systems and the Automation of Energy Systems.'

A. Fedosov and E. Pusenkov (who work at a branch of the open Joint Stock Company (SO UES, Siberia) in their article 'Problems arising during the introduction of microprocessor technology in emergency automation' (in the journal *Power Plants*, 2009 No. 12) note the lack of robust, integrated requirements for the material aspects of DPR and their programming. As a result, there is a very high profusion of programmes and algorithms incorporated into DPR, used in a single energy system, leading to problems during operation, and to an increase in the likelihood of the failure of a given device. D. Rayworth and M.A. Rahim (PB Power, UK) [1.36] also wrote about the dramatic rise in the level of complexity in the work of personnel servicing the relay protection during the transfer from electromechanical to DPR, as well as the reasons behind serious accidents involving energy systems. A. Belyaev, V. Shirokov, and A. Yemelyantsev (who work at the Specialized directorate 'Lenorgenergogaz' in St. Petersburg) also wrote about the complexity of the programming interface and the need to introduce an extraordinary number of settings when programming a DPR, in their article: 'Digital Relay Protection and Automation Terminals. The Practice of their Adaptation to Russian Specifications' (in the journal *Electro-technical News*, 2009, No. 5).

Kovalev B.I., Naumkin I.E. (the Siberian Scientific-Research Institute of Power Engineering); Bordachev A.M., (of the Open joint Stock Company 'Institut Energoset'proyekt'); M. Matveyev and M. Kuznetsov (of the Joint Stock Company 'EZOP'); R. Montignies, B. Jover (Schneider Electric, France); V. Nadein ('Arkhenergo'), V. Lopukhov (Sate Unitary Enterprise 'PEO Tatenergo'); A. Yermishkin (of the Joint Stock Company 'Mosenergo'); R. Borisov (of the Research and Production Company 'ELNA' based in Moscow); A.W. Sowa, J. Wiater (Electrical Department, Bialystok Technical University, Poland) and other specialists also noted the unsatisfactory condition of the electromagnetic environment at the majority of the old substations that were designed and built for electromechanical and not DPR, and the failures that occurred in the operation of DPR as a result. Many of them noted the sensitivity to electromagnetic interference in protective relays on a microprocessor element base was higher by several orders of magnitude than on traditional electromechanical analogues and as such in order to ensure the electromagnetic compatibility (EMC) of secondary circuits the level of electromagnetic protection in these relays needs to be increased dramatically. Without a package of work being carried out to ensure EMC it would be impossible to achieve acceptable reliability levels in DPR.

Closely linked to the lack of stability found in DPR in relation to EMC is another more complex and serious problem with respect to intentional remote destructive electromagnetic threats to DPR, which we first brought to the attention of specialists in [1.36]. Today in many countries across the world equipment has already been developed that is capable of taking industrial microprocessor systems of any kind out of action (which naturally includes DPR). Therefore, not only are many publications in technical journals written by well-known specialists such as Manuel. W. Wik (Defence Materiel Administration, Sweden) and William A. Radasky (of the Metatech Corporation, USA) devoted to this topic, but also reports produced by special commissions under the US Congress (see the 'Report of the Commission to Assess the Threat to the United States from Electromagnetic Pulse Attack', 2008, for example).

Another new problem, which was previously unknown in relay protection is the problem of the cyber vulnerability of DPR (and consequently in the energy system as a whole) to attacks from hackers. Paralysis of the control system and the large-scale disconnection of entire energy systems, chaos in the monitoring system, as well as the disconnection of the Internet and the mobile communications network – according to American proponents these are the likely consequences of a cyber-attack. Moreover, considering the strategic importance of a target like an energy system it would not be lone hackers that would embark on an attack like this but entire military cyber divisions, which have already been set up in many countries around the world. Just last year a separate Cyber division was founded under the National Security Agency (NSA) in the USA, one of the most powerful and highly classified of the world's secret services, and led by General K. Alexander, which brought together all the Pentagon's cyber protection divisions that existed at the time. Some of them will be ensuring the security of not only of the military and state infrastructure, but also the country's most important commercial assets. Understandably a structure of this magnitude will not only be engaged in defending the country from cyber-attacks but will actually develop attacks of its own (the best defence is offence). The head of Cyber Command and the Director of NSA General K. Alexander announced at Congressional Armed Services Committee hearings that cyber warfare has an effect comparable to the use of a weapon of mass destruction. Cyber warfare is developing at great speed. Many countries such as the USA, Russia, China, Israel, UK, Pakistan, India, North and South Korea have developed complex cyber weaponry, which specialists in cybernetics assert are capable of penetrating computer networks more than once and destroying them. In 2010 the cyber budget of the United States was $8 000 000 000 and in the future this is only set to rise. In 2011 the USA is preparing to set a new doctrine on cyber warfare. The direction of this doctrine can be judged by a programme article published in September and written by the Deputy Head of the Pentagon William Lynne III with the symbolic title of 'Defending a New Space'. The basic premise is that from now on the USA will consider cyberspace as potential a battlefield as land, sea and air. Parallel to this NATO began work to create a concept of collective cyber defence. At an alliance summit held in November 2010 it was decided to develop a 'Plan of action in the field of cyber defence'. This document was to have been published by April 2011 but was signed in June. A great deal of emphasis in this document is placed on the creation of a NATO centre to react to cyber incidents. Initially this was due to be launched in 2015 but on the insistence of the United States the deadline was brought forward by three years. The effectiveness of cyber weaponry can be judged by the widely publicized cyber-attack on the Iranian uranium enrichment centre at Natanz with the help of a Win32/Stuxnet computer worm that destroyed hundreds of centrifuges. A further large-scale attack on the Japanese corporation Mitsubishi Heavy Industries, that produces F-15 aircraft, Patriot anti-aircraft missile systems, submarines, surface ships, rocket engines, ballistic missile guidance and interception systems and other military technology occurred in September 2011.

The corporation's computer equipment (45 closed servers and around 50 personal computers) turned out to be infected with a whole range of viruses that had taken control of them completely. This meant that the computers could be controlled remotely and the information on them could be transferred. There were viruses that enabled built in microphones in the computers as well as cameras to be activated. This enabled the plotters to follow what was happening inside the production and research facilities remotely. Some of the viruses erased the traces of the breach, which made any assessment of the scale of the damage much more difficult. Information on the computers that had been taken over was downloaded onto 14 different sites abroad including in China, Hong Kong, the USA and India.

Modern technology enables viruses to be launched into a computer system remotely in the form of enciphered radio waves with the help of pilotless airborne repeater equipment. Wireless Wi-Fi systems, which it is envisaged Smart Grid systems will be based on, are especially vulnerable to these external attacks. DPR manufactured by leading Western manufacturers are already being fitted with built in modems for Wi-Fi.

In the past several attempts were recorded by Iran to penetrate the Israeli energy system. The Senior Analyst of the United States' Central Intelligence Agency (CIA) Tom Donahue announced at a meeting of government officials and employees of American companies that possessed electrical, water, petroleum, and gas distribution systems, of several attempts known to the CIA to penetrate America's energy systems.

It is very obvious that the trends described here will only increase in proportion with the development of this technology, if the following steps to stop these trends are not taken:

1) In the field of the reliability of relay protection:
   1.1. The Introduction of qualified methods of calculating the reliability of DPR [1.37] and a new indicator of their reliability [1.38] that is convenient and practical, and which enables the consumer to submit a complaint to the manufacturer, in place of the current and less than informative indicator 'mean time between failures' (MTBF).
   1.2. Limiting and optimising the number of functions in a single DPR module [1.39, 1.40].
   1.3. A rejection of the use of non-deterministic logic in DPR [1.24, 1.25, 1.41].
   1.4. A considerable restriction on the use of freely programmable logic in DPR – this is a source of human error and of a large number of failures in protective relays [1.24, 1.25, 1.41].
   1.5. The introduction of a ban on using DPR for purposes that do not bear any relation to relay protection, such as monitoring the condition of electrical equipment or for so-called 'protection through preventative action'. [1.24, 1.25, 1.41].
   1.6. A rejection of the use of wireless network technology in protective relays.
   1.7. Forcing DPR manufacturers by law to be concerned with the cyber security of their products and their resilience to intentional destructive electromagnetic threats. This is why standards need to be drawn up and special sections introduced into the technical documentation for DPR, which should reflect the safeguards, and the level of protection in a specific DPR from the attacks

indicated previously, together with its compliance with established standards. Gradual restrictions and then subsequently a complete ban should be introduced on the use of DPR in electrical power engineering that do not conform to the requirements for protection from the threats listed here.

1.8. Publish special bulletins for engineering companies that are engaged in the design of protective relay systems, providing a detailed description of the dangers facing protective relays today, as well as possible preventative measures to protect against them [1.11]. The gradual introduction into design practice of established standards and safeguards initially in newly introduced, and subsequently in existing power systems.

1.9. To assign the development of specific programming and equipment safeguards against the threats listed here to the leading scientific organizations as well as the testing, organization of operational trials and subsequent manufacture of already established, and at the same time inexpensive, equipment safeguards proposed in [1.11, 1.42, 1.43].

2) In the field of the standardization of relay protection:

2.1. Introduce into normative and technical documentation unified definitions for the most important notions in relay protection, such as those proposed in [1.44], for example.

2.2. Develop general technical requirements for microprocessor based protective and automation devices for power systems based on international standards and publish a new document, that uses, for example, the requirements set out in [1.44] as a basis.

2.3. Unify the design as well as the basic software shell for different makes and models of DPR, for which it would be necessary to draw up a set of standards with integrated technical specifications for the design of the functional modules of a DPR, and for the internal communication protocols between them, as well as the basic user software shell.

2.4. Standardize the testing of DPR using modern programmable protective relay testing systems alongside ready-made programming module packages [1.44].

In our opinion, these measures are capable of halting the further development of dangerous trends in the DPR field, and will support a significant increase in the reliability of relay protection, and its resilience to intentional electromagnetic destructive threats as well as reducing running costs.

References

1.1 Popkova N.V. The Philosophy of Technology - the Internet-portal of the Bryansk branch of the Russian Philosophical Society (http://sphil.iipo.tu-bryansk.ru/)

1.2 Vinge V. The coming technological singularity: How to survive in the post-human era. *NASA*. Lewis Research Center, Vision 21: Interdisciplinary Science and Engineering in the Era of Cyberspace pp. 11–22 (SEE N94–27358 07–12), 12/1993.

1.3 Moore G.E. Cramming more components onto integrated circuits - *Electronics,* 19 April, 1965, pp. 114–117.

1.4 Sukharev M. An explosion of complexity - *Computerra,* No. 43, 3 November 1988 (http://offline.computerra.ru/1998/271/1828/).

1.5 Negodayev I.A. The Philosophy of Technology: A textbook/DGTU (textbook)/Rostov on/D, 1998, 319 pp.

1.6 Lopota V.A., Yurevich E.I. Unified mechatronic microsystem modules - the foundation of the intellectual technology of the future - *Artificial Intelligence*, 2002, No. 3, pp. 303–304.

1.7 Bezmenov A.E. *Tolerances, Settings and Technical Metrology. A Textbook for Technical Colleges*. Moskva Mashinostroyenie 1969 322 pp.

1.8 Gurevich V.I. DPR. Equipment, problems, prospects. - M.: *Infra-Inzheneriya*, 2011. - 336 pp.

1.9 Burenok V.M. How can Russia's defensive capabilities be guaranteed in the future? - *The Military Industrial Courier*, No. 39 (507), 9 October 2013.

1.10 Gindiles L.M. Models of civilisations in the SETI problem - *Social Sciences and the Modern World*, 2000, No. 1. pp. 115–123.

1.11 Gurevich V.I. The vulnerabilities of DPR: problems and solutions - M.: *Infra-Inzheneriya*, 2014. 256 pp.

1.12 Grishchuk Yu.S. Timoshenko R.F. An analysis of the reliability of DPR. A Collection of essays 'The NTU Herald': The Problem of Improving Electrical Machinery and Apparatus, No. 16, - *The NTU Herald*, 2010.

1.13 Vnukov A.A. The practice of introducing microprocessor terminals in the modern context - *Electro and Electrical Technology, Electrical Power Engineering, Electro-technical Industry*, 2008, No. 1, pp. 40–41.

1.14 Sapa V.Yu. Electromagnetic compatibility in contemporary electrical power engineering - *Material from the Conference on 'The Achievements of the Higher School. The Technical Sciences'* (the A. Baytursynov Kostanay State University), Kazakhstan, 2011.

1.15 Arynov A.K., Yunus M.E. a comparative analysis of digital protective relays - K.I Satpayev *Kazakh State University Herald*, 2011, No. 83.

1.16 Lint M.G. Mathison V.A., Mikhaylov, A.V. The current state of electro-mechanical protective relay systems and their prospects - *Relay Protection and Automation*, 2013, No. 2, pp. 38–40.

1.17 Kolesnik S.P. The strategic direction of an equipment manufacturer producing power engineering and relay equipment - Abstracts from a Seminar Entitled 'Relay Protection Equipment and Automation, 2013', No. 2, pp. 38–40.

1.18 Grebennikov M. Defining the right path - Russia's Power Engineering and Industry, No. 18 (134) September 2009.

1.19 Iov A.A. and Iov I.A. The reliability of DPR: Myths and Reality - The section entitled 'The Supply of Electrical Power and Electrical Equipment Control Systems in the Mining Industry.' *The All-Russian Scientific and Practical Conference 'The Innovative Development of the Mining and Metallurgical Sector' Irkutsk State Technical University*, Irkutsk 1–2 December 2009.

1.20 Gurevich V.I. Intellectual networks: New prospects or new problems? *The Electro-technical Market*, 2010, No. 6 (part 1); 2011, No 1 (part 2).

1.21 Gurevich V.I. The reliability of DPR: Myths and reality - *Problems in Electrical Power Engineering*, 2008, No. 5–6, pp. 47–62.

1.22 Gurevich V.I. Further thoughts on the reliability of DPR - *Problems in Electrical Power Engineering*, 2009 No. 3 (29), pp. 40–45.

1.23 Gurevich V.I. Is relay protection safe from the point of view of energy security? - *Energy Security and Energy Saving*, 2010, No. 2, pp. 6–8.

1.24 Gurevich V.I. 'The intellectualisation' of relay protection: Good intentions or the road to Hell? - *Electrical Networks and Systems*, 2010, No. 5, pp. 63–67.

1.25 Gurevich V.I. Sensational discoveries in the field of relay protection - *Russia's Power Engineering and Industry*, 2009, No. 23–24, p. 60.

1.26 Gurevich V.I. Logic in free flight - *PRO Electricity*, 2008, No. 1 (25), pp. 28–31.

1.27 Gurevich V.I. Testing DPR - *PRO Electricity*, 2011, No. 2, pp. 28–31.

1.28 Gurevich V.I. Electro-magnetic terrorism - The new reality of the 21st century - *The World of Technology and Engineering*, 2005, No. 12, pp. 14–15.

1.29 Gurevich V.I. The problem of the electromagnetic impact on DPR - *Components and Technology*, 2010, No. 2, pp. 60–64; No. 3, pp. 91–96; No. 4, pp. 46–51.

1.30 Gurevich V.I. The problem of the resilience of DPR and automated systems to intentional destructive electromagnetic threats - *Components and Technology*, 2011, No. 4 (part 1); 2011, No. 5 (part 2).

1.31 Gurevich V.I. The use of cyber weapons against power engineering - *PRO Electricity*, 2011, No. 1, pp. 26–29.

1.32 Gurevich V.I. DPR: New prospects or new problems? - *Electro-technical News*, 2005, No. 6 (36), pp. 26–29.

1.33 Gurevich V.I. On several evaluations of the efficiency and reliability of DPR - *The Electrical Power Engineering News*, 2009, No. 5, pp. 29–32.

1.34 Gurevich V.I. Current problems in relay protection: An alternative view - *The Electro-technical News*, 2010, No. 3, pp. 30–43.

1.35 Gurevich V.I. Criteria for assessing relay protection - Is it worth complicating the situation? - *The Electro-technical News*, 2009, No. 6, pp. 45–48.

1.36 Problems with DPR [in Russian] – Available online at http://digital-relay-problems.tripod.com (accessed August, 2016).

1.37 Gurevich V.I. Problems in assessing the reliability of relay protection - *Electricity*, 2011, No. 2, pp. 28–31.

1.38 Gurevich V.I. New criteria are needed to assess the reliability of DPR - *The Electro-technical Market*, 2011, No. 6, pp. 70–74.

1.39 Gurevich V.I. Current problems of standardisation in the field of relay protection - *The Electro-technical News*, 2012, No. 6, pp. 28–38.

1.40 Gurevich V.I. On the multifunctional protective relay - *PRO Electricity*, 2012, No. 42–43, pp. 45–48.

1.41 Gurevich V.I. Surrealism in relay protection - *EnergoStyle*, 2010, No. 1, pp. 5–7.

1.42 Gurevich V.I. Is protection for relay protection necessary? - *Electrical Energy. Transfer and Distribution*, 2013, No. 2, pp. 94–97.

1.43 Gurevich V.I. Protective relay protection equipment - *Components and Technology*, 2013, No. 5.

1.44 Gurevich V.I. *Problems of Standardisation in Relay Protection*. SPB.: The Publishing House DEAN, 2015, -168 pp.

2

Intentional Destructive Electromagnetic Threats

2.1 Introduction

The use of a special weapon, one capable of destroying an electrical power system as well as other critical elements of the national infrastructure without having a direct effect on human beings, is very attractive in as much as it could lead to the collapse of an entire country. Moreover, those responsible for taking the decision to use such a weapon are not in a position to condemn anyone for the mass murder of the civilian population since it does not have an adverse effect on humans. This weapon encompasses systems that generate very powerful electromagnetic fields that can take electrical apparatus and electronic equipment out of action.

The problem of intentional destructive electromagnetic threats has, in recent years, become more and more topical in connection with two contemporary trends: the expanding use of microelectronic and microprocessor technology in electrical power engineering on the one hand, and the intensive development of remote weapons capable of targeting electronic apparatus on the other [2.1]. Furthermore, this problem is not limited to such a profoundly civilian sector as electrical power engineering but also the military in as much as the military bases and ranges receive electrical power and water from civilian systems, and any serious failures in the operation of these systems inevitably takes its toll on the military, even though all their weapons systems are protected from intentional destructive electromagnetic threats.

2.2 A Brief Historical Background

The destructive effect of a remote nuclear blast on electronic apparatus was discovered during the very first tests with what was then a new type of weapon. The theoretical basis for this phenomenon by which powerful electromagnetic pulses are formed was discovered as a consequence of the theoretical writings of Laureate of the Nobel Prize for Physics, Arthur Compton, which he conducted back in 1922. The military were quick to assess the prospects for using this phenomenon as a weapon designed to strike enemy infrastructure, first and foremost electricity supply systems. The first direct experiments to study HEMP were conducted on 9 July 1962 by the Atomic Energy Committee and the Nuclear Safety Agency of the Department of Defence of the United States of America (this project was given the codename 'Starfish Prime'). A missile with

Protection of Substation Critical Equipment Against Intentional Electromagnetic Threats,
First Edition. Vladimir Gurevich.

a 1.44 megaton thermonuclear warhead was launched from a US test-range located on Johnson Attol between the Marshall and Hawaiian Islands in the Pacific Ocean to an altitude of around 450 km and would then be detonated. This test was just one of five high altitude nuclear blasts aimed at studying HEMP that were carried out by the United States in 1962 within the confines of a wider project codenamed 'Operation Fishbowl'. During the course of these tests powerful electromagnetic pulses were recorded that had a devastating effect on electronic equipment, communication and overhead power lines (OPL), radio and radar stations and even took the street lighting out of action on the Hawaiian Islands, located some 1500 km from the epicentre of the explosion [2.2].

In that same year, 1962 (on 22 October, 28 October and 1 November), a series of three high altitude nuclear tests were conducted in the Soviet Union within the confines of the so-called 'Project K', each with a yield of 300 kilotons (K3–184, K4–187 and K5–195) with the aim of studying HEMP. Missiles fitted with nuclear warheads were launched from the missile testing range at Kapustin Yar in the Astrakhan region and detonated at altitudes ranging from 60–290 km above the territory of the military range at Sary-Shagan in the Karaganda region in Kazakhstan (the Closed Administrative Territorial Unit Priozersk). Work on HEMP research and to prepare for these high altitude tests was conducted in the USSR by the Central Institute of Technical Physics of the Department of Defence (either 'Unit 51105' or the Central Scientific Research Institute-12) in Sergiyev Posad in the Moscow region (today this is known as the Federal State Institution '12th Central Scientific Research Institute of the Department of Defence of the Russian Federation'). During the course of one of the tests (K3–184) pulse currents of up to 3400 A were recorded in overhead telephone cables, which was brought about by the advent of a pulse voltage amplified up to 28 kV, the activation of all the discharge devices fitted to this apparatus as well as the failures of all the fuses was recorded, and this was accompanied by the failure of the communication system, as well as damage to radio communication systems at a distance of 600 km from the epicentre of the explosion, a radar station was put out of action that was located a distance of 1000 km away, transformers and generators at power stations were damaged, and the isolators in the power lines (OPLs) flashed over, see Fig. 2.1. Serious damage to apparatus was recorded at the Baykonur cosmodrome. Furthermore, this relates to apparatus dating from the 1960s that was manufactured using electro-mechanical elements and radio valves, which are much more resilient to the effects of HEMP than modern micro-electronics and microprocessor technology.

Apart from that, in both the American and Soviet tests thermonuclear warheads were used, for which the electromagnetic pulse would be 3–5 times weaker than that produced on detonation of a standard nuclear warhead of a similar yield.

2.3 The First Reliable Information on HEMP as Well as Protection Methods in the Field of Electrical Power Engineering

It is perfectly natural that in view of the complexity, importance and the high cost of conducting nuclear tests all the information on these tests be meticulously classified and that the first to own this information would be military specialists. It could be

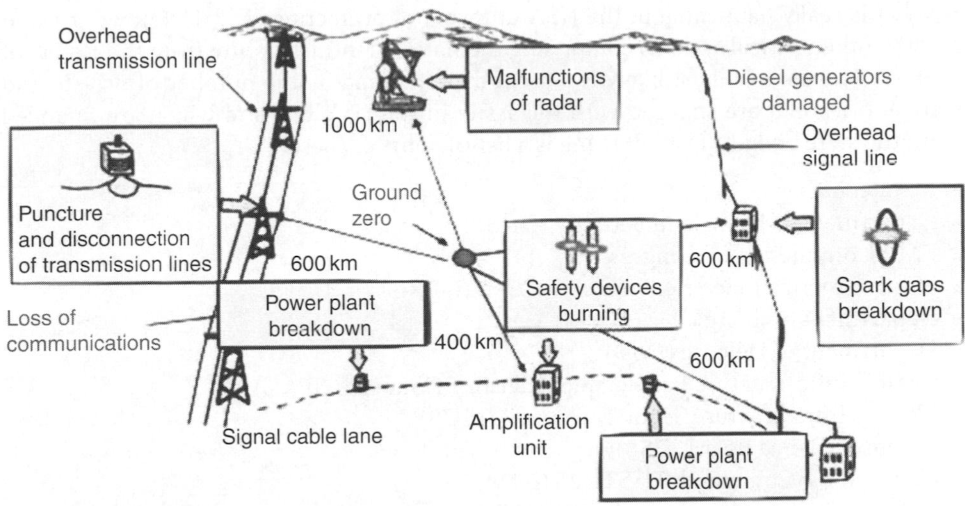

Fig. 2.1 An illustration of the damage to equipment sustained as a result of the effects of high altitude HEMP over Kazakhstan in 1962. This diagram was initially presented to the head of Central Scientific Research Institute-12, Major-General and Doctor of Technical Sciences Professor V.M. Loborev in English at the international EUROEM conference in France in 1994 [2.3].

suggested that the first information of this nature to come to light was presented during *perestroika* by the Head of the Central Institute of Technical Physics of the Department of Defence, Major General V.M. Loborev in his well-known report at the EUROEM conference in France in 1994, see Fig. 2.1. This, however, is not the case at all. It appears that the first publications in the open mass-media to contain detailed and reliable information on HEMP parameters and its influence on the country's infrastructure, and specifically on the electrical power supply system, date back to the end of the 1960s and the beginning of the 1970s– that is to say all this information has been available to the public for 40–50 years [2.4–2.23]. Furthermore, some of these publications such as in [2.16] and [2.18] contain a detailed description of measures to protect electrical equipment from the effects of HEMP. Evidently most of these publications were published in the United States of America and therefore it would be logical to suppose that in the course of the last 50 years the United States would have achieved unparalleled success in protecting the most important components of her national infrastructure from the effects of HEMP. Furthermore, the Army should have an interest in this.

2.4 The Actual Situation with Respect to the Protection of Power Electrical Systems from HEMP and other Types of Intentional Destructive Electromagnetic Threats

> *You can fool all of the people some of the time; you can even fool some of the people all of the time, but you cannot fool all of the people all of the time.*
>
> *Abraham Lincoln*

What is really happening in the USA in terms of protecting electrical power system and the other critical systems comprising the national infrastructure from the effects of HPEM? The answer most likely is a great deal, judging by the number of private and state entities that are engaged with this issue in the US alone and which are financed from the state budget. The following is a list of some of these:

- Metatech Corp.
- Department of Homeland Security (DHS)
- EMP Commission of Congress
- North American Electric Reliability Corp. (NERC)
- Department of Energy
- Department of Defense (DoD)
- Critical Infrastructure Partnership Advisory Council (CIPAC)
- Electric Infrastructure Security Council (EICS)
- Defense Science Board (DSB)
- US Strategic Command (USSTRATCOM)
- Defense Threat Reduction Agency (DTRA)
- Defence Logistics Agency (DLA)
- Air Force Weapons Laboratory
- FBI
- Sandia National Laboratories
- Lawrence Livermore National Laboratory (LINL)
- Oak Ridge National Laboratory
- Idaho National Laboratories
- Los Alamos National Laboratories
- Martin Marietta Energy Systems, Inc.
- National Security Telecommunications Advisory Committee
- Federal Emergency Management Agency (FEMA)
- National Academy of Science
- Task Force on National and Homeland Security
- EMPrimus
- Neighborhood of Alternative Homes (NOAH)
- EMPact America
- Federal Energy Regulatory Commission (FERC)
- Electric Power Research Institute (EPRI)
- NASA
- U.S. Northern Command (NORTHCOM)
- SHIELD Act
- EMP Grid
- EMP Technology Holding
- Strategic National Risk Assessment (SNRA)
- Walpole Fire Department

International organizations participating alongside the USA:

- International Electrotechnical Commission (IEC), Technical Sub-committee 77C
- CIGRE, Working Group WG C4.206.

Dear reader, do you not find the active participation of such a large number of organizations in just one country in this field, one which an enormous amount of material has been published for the past few decades and which practically no stone has been left unturned that would require further research slightly suspicious?

It appears that intentional destructive electromagnetic threats, and specifically HEMP, is nothing less than a wonderful and long standing tool for chipping away at the state budget. Similarly, nobody has expressed any interest in the notion that the process of 'chipping away' at the budget should ultimately culminate in concrete action of some kind being taken to protect electrical power systems. A quote by one of the former officers under the Department of Defence of the United States of America Dr Ashton Carter serves as confirmation of this: 'The Army, Navy, and Strategic Command continue to think that they need to think about the problem.' The Executive Director of the Task Force on National and Homeland Security Dr Peter Vincent Pry was more specific on this: 'The problem is not the technology. We know how to protect against it. It's not the money, it doesn't cost that much. The problem is the politics. It always seems to be the politics that gets in the way.'

In his large volume book entitled *Apocalypse Unknown*, see Fig. 2.2, Dr Pry argues that in certain other countries (Israel, UK and Russia) the situation is much better than in the United States and that in these countries work has already begun to realize practical steps to protect electrical systems. We should hurry to reassure Dr Pry: he does not need to worry about America's backwardness. The situation in this field, in Russia for example, was actually much worse than in America, in as much as Russian specialists in electrical power engineering have either not heard of this problem at all, or considered it Gurevich's 'horror story' (in as much as the only author writing about this topic in Russian language publications, is the author of this book). The situation in other

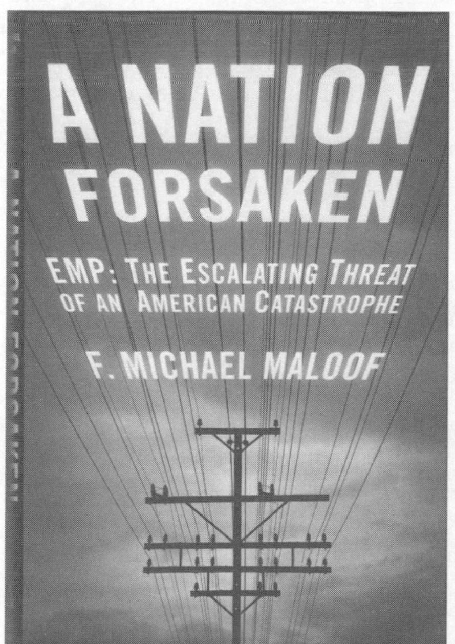

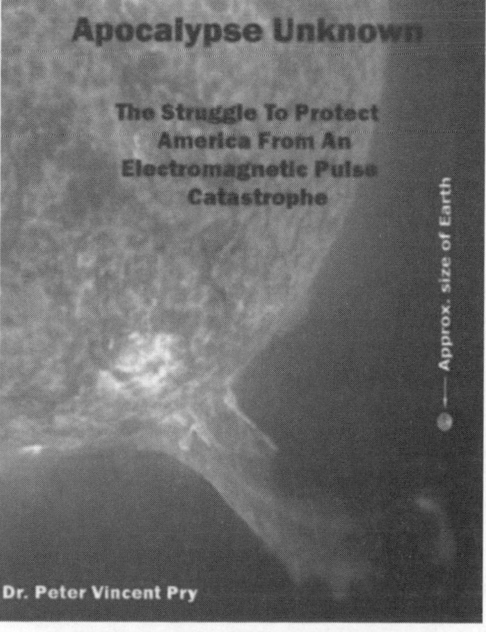

Fig. 2.2 Michael Maloof's book *A Nation Forsaken*, which is devoted to describing the bureaucratic and political games around the problem of intentional destructive electromagnetic threats in the United States of America (on the left) and Dr Peter Vincent's Pry's book *Apocalypse Unknown* (on the right).

countries was no better. In general, it became clear why over the course of decades absolutely nothing of any consequence has been done to protect electrical power systems from HPEM and what has been done is limited to research results, and reports running into many pages, seminars, conferences and other pleasant pastimes amongst a circle of colleagues. Put simply the many 'participants in the process' are not interested at all in the culmination of many years of research, but are interested in keeping this topic 'afloat' and in its ongoing finance.

A book written by the former Pentagon Analyst Michael Maloof, *A Nation Forsaken* (see Fig. 2.2), is dedicated to this problem. The aforementioned Peter Vincent Pry also writes about the serious bureaucratic obstacles surrounding this issue in his book *Apocalypse Unknown*.

The representatives of the powerful Military-Industrial Complex (MIC) also did their bit to slow down the process of implementing concrete measures, which were already well known, to protect systems from intentional destructive electromagnetic threats. They insist that the only effective defence against HEMP is a national anti-missile defence system, into which more budgetary funds should be invested. This position, which was adopted by the representatives of the MIC, would be completely understandable if the relatively low cost of defensive measures to protect the most important elements of the country's infrastructure and its systems from HEMP were compared with the cost of developing and producing an effective multi-layered anti-missile shield that would protect the entire country. As far as the other, non-nuclear intentional destructive electromagnetic threats are concerned [2.1] it is possible to conclude that they simply do not exist in as much as an anti-missile defence system would not protect against them and that information on these threats, which is constantly being published in the media, is no more than bluff aimed at scaring housewives. However, it appears that it is not that simple and that missile systems have been in existence for some time that an anti-missile system is not capable of defending against, that is to say it is not possible to protect the national infrastructure from HEMP attacks. What sort of systems are these then?

2.5 Medium and Short-Range Missile Systems – Potential Sources of Intentional Destructive Electromagnetic Threats that Anti-Missile Defence Systems Are Powerless to Defend Against

There is an observable trend today, in which the yields of all types of nuclear warheads are being reduced in connection with improvements in their accuracy. Thus for example if the Russian 'Tochka-U' (9M79B2), which has a circular error probable (CEP) of 250 m is fitted with a nuclear warhead with a yield of up to 200 kt (AA-92 charge type) then the much more accurate and newer Russian 'Iskander' (9 M723) missile, see Fig. 2.4 later, which has a CEP of up to 30 m is fitted with a nuclear warhead with a yield of just 50 kt, see Fig. 2.3. A yield of 50 kt is simply not enough however to create a powerful and effective HEMP.

The Russian 'Iskander' system, which was widely publicized as having no analogues, appears in point of fact to be not so unique. The Israeli LORA (LOng Range Attack)

(a)

(b)

Fig. 2.3 Launchers carrying the 'Tochka-U' tactical missile (a) and the 'Iskander' short-range attack missile (b).

missiles demonstrate very similar tactical performance characteristics, likewise aspects of their trajectory, and control systems. Moreover, they demonstrate greater accuracy when compared to the 'Iskander' (their CEP is 10 m), the weight of the missile is half that of the Iskander and the warhead is more powerful. They are capable of carrying a more powerful nuclear charge and use a universal launching system that can be loaded onto different modes of transport, including ships.

The LORA launching system is manufactured in the form of a container with four missiles, which in terms of their shape are very reminiscent of the containers used in the Russian 'Club-K' with the same number of 3 M-14KE, Kh-35UE missiles, see Fig. 2.4.

The Club-K is the Russian container based missile weapons system, which can be housed in a standard 20 or 40-foot sea container.

This system is designed to strike surface as well as ground targets. These systems can be installed on coastlines, as well as on different classes of vessels, and flatbed rail and road wagons. They can be fitted with anti-ship missiles (such as the 3 M-54KE, 3 M-54KEI and the Kh-35UE) or with missiles designed to strike ground targets. All the missiles used in this system are cruise missiles, which fly at a low altitude between 10–150 m and are not designed to be fitted with nuclear warheads while the Israeli container based system LORA is equipped with short range attack missiles that climb to an altitude of up to 45 km and are capable of delivering high yield nuclear charges at a distance of up to 300 km.

Fig. 2.4 The container based launching systems for the Club-K (top) and LORA (below) missile systems.

Why is it that these missile systems specifically are being examined in detail? The reason is that it is these same systems, with relatively small missiles housed in standard sea containers on ships close to the coast or actually in port (Fig. 2.5) and which are capable of delivering nuclear charges at a range of hundreds of kilometres, and of climbing to an altitude of dozens of kilometres that represent sources of HEMP, and are invulnerable to any anti-missile defence systems, both existing, and prospective, in view of their covert approach to the target, their exceptionally short flight time, and their trajectory, which can be altered in flight.

The ability to covertly bring short-range tactical missiles fitted with nuclear warheads close to a target, to rule out the possibility of their being hit by anti-missile defence systems on the one hand, and on the other to ensure that they are not covered by restrictions imposed by international treaties has long been understood by specialists, and attempts began to be made to create these systems immediately after the advent of tactical missiles that were small in comparison and which were fitted with nuclear warheads. Thus the 'Little John' (MGR-3) missile system that employed unguided missiles that were capable of carrying nuclear warheads entered service with US paratrooper units in 1961 in the United States of America. The lightweight launching systems for these missiles were transported in CH-47 'Chinook' helicopters both in the cabin, and on external hard points.

Fig. 2.5 Containers on ships and in port in which short-range tactical missile systems fitted with nuclear warheads can be stored and which are invulnerable to anti-missile defence systems. *Sources:* top figure, © Hans Hillewaert. https://commons.wikimedia.org/wiki/File%3AElly_Maersk.jpg. Used under CC-BY-SA 4.0 https://creativecommons.org/licenses/by-sa/4.0/deed.en; Bottom figure, National Oceanic and Atmospheric Administration.

The Soviet Union was quick to assess the potential of systems such as these and in accordance with top secret decree No. 135-66ss dated 5 February 1962 issued by the Council of Ministers work began to create a tactical missile system the 'Luna-MV' (9K53) based on the 9M21B missile fitted with a nuclear warhead or 9M21B1 thermonuclear warhead and a 9P114 launching system, which was a light, self-propelled platform with an M-407 carburettor engine producing 45 hp taken from a 'Moskvich' car. Several modifications of these missile systems were subsequently developed, which were designed to be transported by cargo carrying Mi-6 or Mi-10 helicopters. It was envisaged that the helicopter would be able to deliver the missile along with the launch system behind enemy lines. Then the system would be able when necessary to travel further on its own chassis and then suddenly attack using the missile from a location where the enemy would not anticipate a missile system to be, which in actual fact turns a tactical system into a strategic one. Work on the 'Luna-MV' system reached the prototype testing stage. However, quite a few problems were encountered, including the helicopter's large 'sail area' with the launch system suspended underneath and correspondingly its considerable wind drift, as well as the insufficient range of the helicopters when loaded. As a result, in 1965 work on this system was stopped.

Contemporary technology has enabled a return to this idea and it has been realized successfully. Today there are hundreds of millions of sea containers in circulation across the world, see Fig. 2.5. Nobody knows which of them are genuine and which are filled with missiles... Despite the fact that today the Israeli LORA system is actually the only fully-fledged container system that is capable of approaching a country's coastline covertly on a container-carrying vessel and of striking her territory with an electromagnetic pulse, the fact that this system does exist enables an affirmation that the assurance made my members of the MIC that advanced anti-missile defence systems alone can provide a reliable defence from HEMP and therefore funds should be invested in these systems alone do not correspond to reality and, in essence, undermine public opinion. The actual situation is such that the Army is not in a position to provide a sufficiently reliable defence of the power systems supplying cities and population centres from intentional destructive electromagnetic threats and as such it is the electrical engineering specialists themselves that need to be concerned with this defence ahead of time.

2.6 What is Needed to Actually Defend the Country Against an 'Electromagnetic Armageddon'?

In as much as all the necessary fundamental research into this problem has already been conducted some time ago and that the results and the practical recommendations have been published in universally accessible sources of information [2.24–2.34] and also in the multitude of International Electrotechnical Commission (IEC) standards [2.34–2.41], as well as those of the institute of Electrical and Electronics Engineers (IEEE) [2.42] and the military standards of the United States Department of Defence [2.43–2.49], it follows that the financing of a huge number of organizations working this problem and which are using it as their main source of income should be stopped and the released funds be directed towards conducting some very concrete activities surrounding the defence of electromagnetic systems from HPEM [2.50]. In those countries where a multidivisional network of organizations working this problem, as in the USA for example, has not been established, the example of the USA in starting to create a similar structure should not be taken since this would lead to a dead end. The only organization that should remain and which should manage the process should in our opinion be the National Coordination Centre for the problem of HPEM, which is tasked with analysing the work produced thus far on this problem, compiling a specific plan with deadlines and organizations that are responsible for meeting those deadlines, to issue concrete technical proposals to these organizations for the protection of electrical supply systems from intentional destructive electromagnetic threats and then organize and coordinate this work. The end result of the activities of this Centre should not be reports or conferences (which should simply be banned!) but actual power stations and substations that are protected from HPEM.

2.7 The Classification and Specifics of High Power Electromagnetic Threats

In English language technical literature, intentional destructive electromagnetic threats are defined as 'High Power Electromagnetic Threats' (HPEM) and are subdivided into two types: a High-altitude Electromagnetic Pulse (HEMP) and Intentional Electromagnetic Interference (IEMI).

An HEMP is a very powerful electromagnetic pulse that is a consequence of a high-altitude nuclear explosion. This powerful electromagnetic pulse that occurs during a nuclear explosion has long been known about as one of the adverse effects of such an explosion. The fact that a nuclear explosion would always be accompanied by electromagnetic radiation arose from the theoretical research work into the effects of X-ray radiation conducted by the American theoretical physicist Arthur Compton back in 1922 (in 1927 he was awarded the Nobel Prize for his discovery). Not much attention was paid to this effect in those far off days and it was recalled with the onset of nuclear testing. In [2.51] this is described as follows:

> At the end of 1946 in the Bikini Attol region (in the Marshall Islands) under the codename "Operation Crossroads" nuclear tests were conducted in which the destructive effects of an atomic weapon were tested. During the course of these nuclear tests a new phenomenon in physics was discovered – the formation of a high power Electromagnetic Pulse (EMP), which immediately attracted great interest. The EMP in high-altitude explosions was particularly significant. In the summer of 1958 nuclear tests were conducted at high altitudes. The first series of tests under the codename "Hardtack" were conducted above the Pacific Ocean close to Johnstone Island.

During the course of these tests two charges were detonated that were in the megaton class: the 'Teak' at an altitude of 77 km and 'Orange' at an altitude of 43 km. In 1962 these high-altitude tests continued: at an altitude of 450 km under the codename 'Starfish' an explosion was conducted of a warhead with a yield of 1.4 megatons. The Soviet Union also conducted a series of tests in 1961–1962 during which research was conducted into the effects of high-altitude nuclear explosions (180–300 km) on the operation of anti-missile defence system apparatus. Powerful EMPs were recorded during the course of these tests, which had a considerable destructive effect on electronic apparatus, communication lines, and on the supply of electricity, as well as radio and radar stations over great distances.

The relationship between the effective impact zone in which electronic equipment was damaged and the altitude at which a charge was detonated with a yield of 10 megatons is set out in Table 2.1.

Table 2.1 The electromagnetic impact zone from a high-altitude nuclear explosion.

Altitude of the explosion, km	The approximate diameter of the impact zone, km
40	1424
50	1592
100	2.242
200	3.152
300	3.836
400	4.402

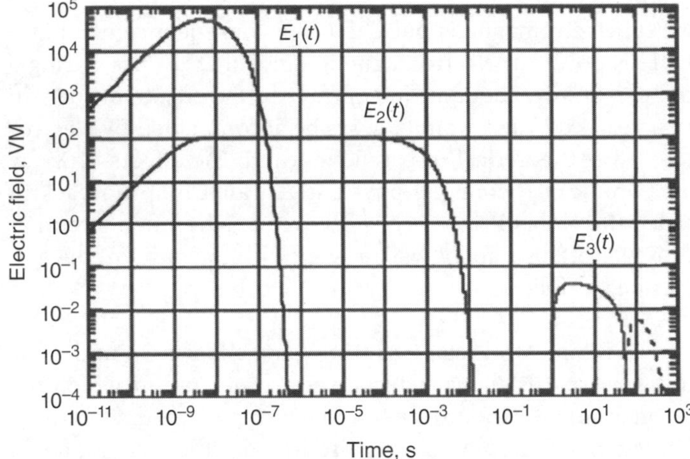

Fig. 2.6 The parameters of the components of a high-altitude nuclear explosion (IEC 61000-2-9).

In accordance with the IEC's classification the three components of HEMP are defined as: E1, E2 and E3, see Fig. 2.6.

E1 is the 'quickest' and the 'shortest' component of HEMP, and is brought about by a powerful flow of Compton's high energy electrons (this is a product of the interaction of γ – the instant radiation quanta of a nuclear explosion with the atoms present in the gases in the air) moving in the Earth's magnetic field at close to the speed of light.

It is this interaction of very fast moving, negatively charged electrons with a magnetic field that produces a pulse of electromagnetic energy, concentrated by the Earth's magnetic field and directed downwards from a certain altitude towards the Earth. The range of this pulse usually grows until it reaches its maximum in the course of 5 nanoseconds and reduces by a half in the course of 200 nanoseconds. According to the IEC's definition the full duration of the E1 pulse could be 1 microsecond (1000 ns). The E1 component is brought about by the most intensive electromagnetic field, that causes high voltages in electrical circuits, it creates in high temperate latitudes pulse voltages close to the ground of up to 50 kV/m at a power density of 6.6 MW per square metre. The E1 component is responsible for the majority of the damage to electronic

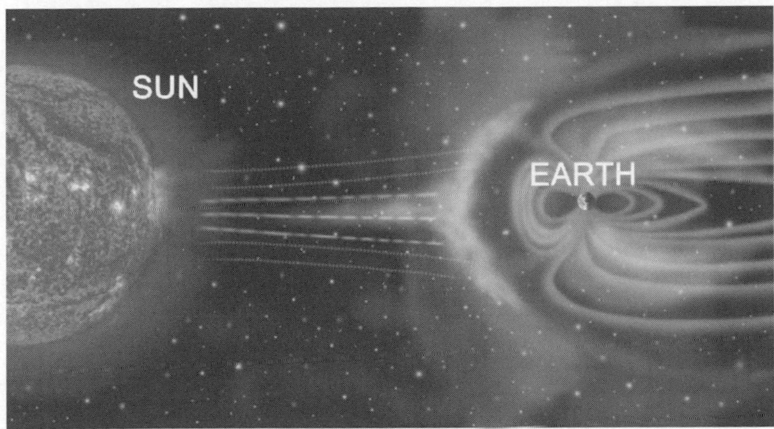

Fig. 2.7 The distortion of the Earth's magnetic field under the influence of emissions of solar plasma.

equipment, which is linked to the effect of overvoltage and electrical flashovers in p-n-junctions in semiconductor elements and insulators. Standard protection elements (zinc-oxide varistors, gas discharge tubes), which are effective in protecting against atmospheric overvoltage, do not always work properly to protect equipment given the effects of the E1 component, and the power they disseminate is far from sufficient to absorb the energy of an E1 component pulse, as a result of which standard protection elements can simply be destroyed.

E2 in terms of its build up and duration is an 'intermediate' EMP component, which according to the IEC definition lasts for approximately 1–100 ms. The E2 component has much in common with the electromagnetic pulses produced in the atmosphere (such as lightning). The voltage of this field can reach 100 kV/m. Owing to the similarities of the parameters of the E2 component to lightning and the well-developed technology to protect against lightning, protection against the E2 element is not considered a problem. However, when the effects of the E1 and E2 components are combined another problem emerges, if the protective elements are destroyed under the influence of the E1 component, after which the E2 component is able to penetrate the apparatus unhindered.

The *E3* component is very different from the other two principle components of EMP. This is a very 'slow' pulse, which lasts ten hundredths of a second and is caused by the displacement and subsequent restoration of the Earth's magnetic field. The E3 component is similar to a geomagnetic storm, generated by a very intensive solar flare. Geo-magnetically induced currents are currents that flow within the Earth and which are caused by geomagnetic disturbances in the Earth's magnetosphere. These currents also vector through long distance metallic objects found in the earth such as pipelines, railway lines and cables. The voltage of an induced field can reach up to 1 V/km. Powerful disturbances in the Earth's magnetosphere are triggered during solar storms, which are accompanied by emissions of an enormous quantity of ionized plasma towards the Earth, see Fig. 2.7.

Electrical currents always flow in the ionosphere, located several hundred kilometres above the surface of the Earth, under the influence of the Earth's magnetic field and of its turning on its own axis. They are supported by the constant formation of a large number of charged particles – ions and free electrons from the molecules of

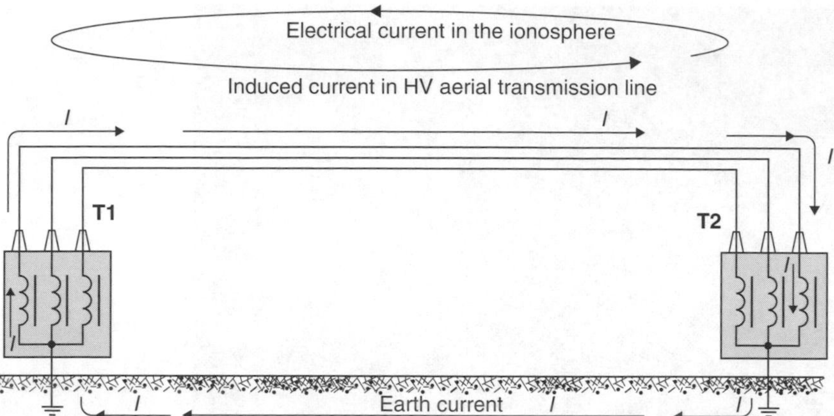

Fig. 2.8 A diagram showing how both OPL and terrestrial currents are vectored by electrical currents in the ionosphere.

atmospheric gases broken down by solar radiation. These electrical currents have a significant influence on the formation of the Earth's magnetic field. During solar storms particularly powerful proton streams and solar plasma electrons sharply increase electrical currents, flowing in the ionosphere.

Dramatic changes in these currents lead not only to dramatic changes in the Earth's magnetic field but also to the creation of geo-magnetically induced currents and the vectoring of powerful currents in long distance OPLs. These vectored currents are connected via the earthed neutrals found in high-power transformers, see Fig. 2.8. Since these currents have a very low frequency the fact that they flow through the armatures of power transformers leads to a saturation of the magnetic circuits in the transformers and to a dramatic reduction in their impedance. As is widely known the constant component in a power transformer current also manifests itself when the transformer is switched on, therefore the protective relays in power transformers are usually tuned out of the constant component and do not react to it. Apart from that direct current (or a very low frequency current) barely makes it through the current transformers. Thus standard relay protection would not react to induced currents saturating the transformer and it would simply burn out. Cases have been documented in history in which power transformers have burnt out under the influence of induced currents during solar storms. Thus in 1989, a solar storm, which was modest in scale, caused very high voltage power transformers to be damaged and the Canadian province of Quebec was plunged into darkness for a period of 9 hours.

A superpower transformer failed at the same time at the Salem Nuclear Power Plant in the American state of New Jersey. On 29 April 1994 a powerful very high voltage transformer failed completely at the Maine Yankee Nuclear Power Plant shortly after the onset of a powerful geomagnetic storm. On 24 March 1940 the supply of electricity in several regions in the states of New England, New York, Pennsylvania, Minnesota, Quebec and Ontario was temporarily disrupted following the most powerful of geomagnetic storms, as well as 80% of all the principle telephone networks in Minneapolis [2.52]. An extraordinarily strong solar storm was predicted by NASA scientists in 2012 (according to some forecasts this was due in 2013). According to forecasts [2.52] strong magnetic storms are possible periodically over the next few years as a result of which failures in energy systems are expected across the globe. These could last from several hours to several months (in connection with the lack of back-up power transformers in many energy systems). This threatens a real collapse of contemporary civilization that is highly dependent on modern technology and is therefore vulnerable to catastrophes of this nature.

The E3 component of a high-altitude nuclear blast can have a similar, in terms of its physical nature, effect on high-power transformers [2.52]. The idiosyncrasies of high-power transformers are such that they cannot be replaced quickly after they have failed, in contrast to electronic devices that also incur damage when exposed to these effects. Given this need for a solution to the problem of protecting power transformers from being damaged by the effects of induced low frequency geomagnetic currents becomes clear.

Since the 1980s in the last century a number of countries across the world have been working strenuously on the creation of a so-called 'Super-EMP' nuclear charge with an enhanced emission of electromagnetic radiation. This work is in the main being conducted in two directions: by means of creating an enclosure around the charge made of a substance that emits high-energy γ radiation when exposed to neutrons from a nuclear explosion, and by means of focusing γ radiation. According to calculations carried out by specialists it would be possible to create a field density close to the Earth's surface in the region of hundreds or even thousands of kW/m with the help of a Super-EMP. Moreover, the military are making no secret of the fact that the main target for these EMP weapons in future conflicts would be government and state administration systems, and the national infrastructure, including electricity, water supply and communications systems.

In June 1950 the Central Institute of Technical Physics of the Department of Defence of the Russian Federation was even created as part of the 12th Central Directorate of the Department of Defence of the USSR in Sergiyev Posad-7 (in the 'Ferma' settlement). This institute was given the number Military Unit 51105 (today this is known as the Federal State Institution 'The 12th Central Scientific Research Institute of the Department of Defence of the Russian Federation') and was headed by the academic Vladimir M. Loborev (since 2002 Rear Admiral, Doctor of Technical Sciences, and Professor Sergey F. Pertsev has been the head of the institute).

The principle task for this institute was to research the adverse effects of a nuclear explosion, principally the electromagnetic pulse but also the laser and particle beam pulse, as well as the ultra-high frequency (UHF) of the weapon, X-ray radiation and so on. Amongst the institute's experimental devices, the GIN-10 very high power high voltage pulse generators are widely known; as are the EMPI-B and EMPI-BMs that imitated an electromagnetic pulse from a nuclear explosion; the 'Arterit' and 'Zenit' systems, which were designed to test how technology reacted to the effects of an electromagnetic pulse, and the 'BARS' pulse nuclear reactor among others.

A nuclear explosion is not the only way to create a powerful EMP. Contemporary achievements in the field of non-nuclear EMP generators enable them to be made sufficiently compact so that they can be used in standard, and highly accurate delivery systems. Therefore, the issue of resilience to the effects of EMP will remain the focus of attention for specialists no matter what the outcome of negotiations on nuclear disarmament.

Intentional Electromagnetic Interference (IEMI) is the second HPEM category and is not linked to a nuclear explosion.

The first theoretical ideas on the possibility of creating non-nuclear superpower EMP explosively pumped flux compression generators were expressed in 1951 by the Academician Andrey Sakharov during his work on a nuclear warhead in Arzamas-16 (today this is known as the All-Russian Scientific Research Institute of Experimental Physics RFNC-VNIIEF).

Robert Lobarev began the first experimental work at this institute on how to obtain superpower pulse magnetic fields by means of their explosive compression, and in 1952 he succeeded in obtaining pulse magnetic fields of 1 500 000 Gauss (by way of a comparison the Earth's magnetic field is only around 0.43 Gauss on the equator and 0.7 Gauss in the polar regions). Subsequently Aleksandr Pavlovskiy and Vladimir Chernyshev, both of whom worked at this institute, continued this research. This collective led by A. Pavlovskiy succeeded in constructing an explosive generator, see Fig. 2.9. with a pulse current of 200 000 000 A, generated in a magnetic field of 10 000 000 Gauss.

A superpower EMP explosively pumped flux compression generator is a ring made from an explosive that surrounds a copper coil. A set of detonators that are detonated

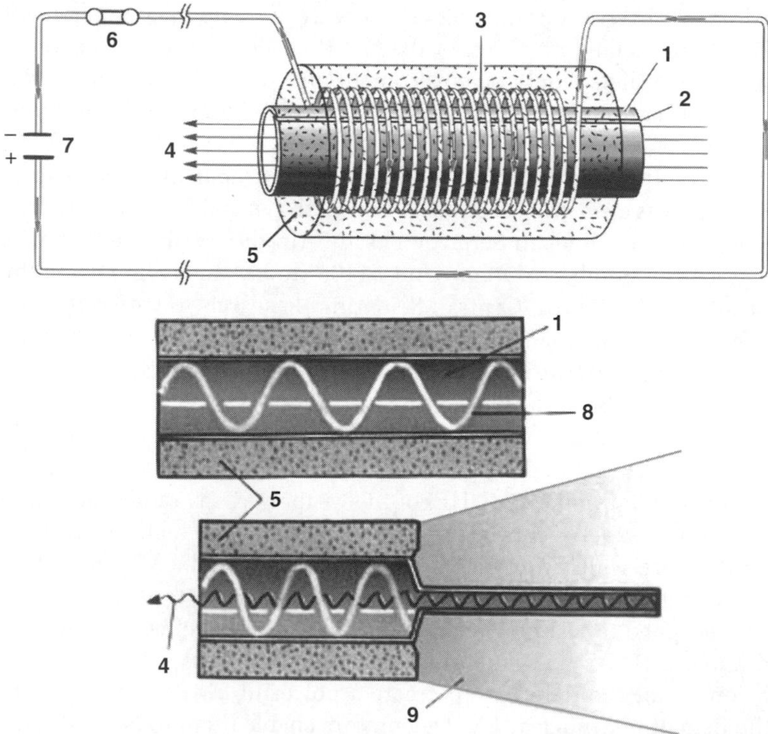

Fig. 2.9 An explosively pumped flux compression generator: 1 – Electromagnetic resonator; 2 – groove; 3 – coil, streamlined by current; 4 – directed electromagnetic radiation; 5 – explosive; 6 – commutator; 7 – energy storage device (capacitor); 8 – standing wave; 9 – debris flying out from the explosion.

synchronically initiate the detonation, which is axipetal. At a point synchronized with the detonation the powerful capacitor would discharge, and the current would form a magnetic field inside the coil. The blast wave under high pressure (around 1 000 000 atmospheres) compresses and 'bridges' the windings in the coil, turning them into a tube and enclosing this field within it. The flow loop is compressed at a speed of several kilometres a second, depending on the type of explosive. As the laws of physics dictate the intensity of a magnetic field created by a loop in this case is proportional to variations in the timing of the rate of inductance. Since the size of the coil changes at very high speed during its collapse then correspondingly the range of the magnetic flux would also be huge (tens of millions of amps). At that moment one of the rings in the resonator is destroyed by an explosive charge and the blast wave, converging at a point and bouncing back, is deflected backwards, changing the field in this single jump. In this case the standing wave becomes a travelling one, and in doing so develops enormous pulse power, which leads to the generation of a pulse flow of radio-frequency electromagnetic radiation. In the space of nanoseconds, the field changes but not in accordance with the law of sinus with a period, equal to the time taken for compression and rarefaction, but more dramatically, and this means that a number of frequencies are present within the function that describes this change in field. Therefore, the source of the explosively pumped flux compression is an ultra-wide band source that is emitted across a range from hundreds of MHz to hundreds of GHz given a pulse duration of tens of hundreds of microseconds.

At almost the same time and independently from the Soviet engineers these same pulses were being obtained by a group led by Max Fowler at the Los Alamos National Laboratory in the United States of America. These scientists met for the first time in Novosibirsk in 1982 at an international conference on super power magnetic fields, and in 1989 a group of Soviet scientists led by A. Pavlovskiy visited the Los Alamos Laboratory.

From almost the very beginning of this work it became clear not only to the scientists in the USA and the USSR but also to the politicians that these sources of super high power electromagnetic pulses could become the basis for the creation of a new type of weapon. The speech made by Nikita Khrushchev in the 1960s in which he made reference to a 'fantastic weapon' of some kind that was being developed by Soviet scientists is testimony to this.

It was the head of the Special Purpose Ammunition Laboratory at the Central Scientific Research Institute of Chemistry and Mechanics Doctor of Technical Sciences Alexander Prishchepenko who made the first official declaration concerning explosively pumped flux compression generators as an independent device that can create very powerful electromagnetic pulses and could be used as a weapon, following successful tests on 2 March 1983 at the Soviet Army's Scientific Research Institute Geodeziya's test-range based in the Moscow region (which today is known as the Federal Treasury Enterprise: the Scientific Research Institute 'Geodeziya').

Subsequently, Associate Member of the Academy of Military Sciences, and Doctor of Technical Sciences A. Prishchepenko formulated the basic principles for the deployment of electromagnetic ammunition in combat. Today, intensive research in the field of IEMI is carried out in several areas and non-nuclear explosively pumped flux compression generators are no longer the only type of non-nuclear electromagnetic weapon. There is a wide range of high power microwave devices. Relativistic klystron tubes and magnetrons, reflex triodes, backward-wave tubes (BWT), gyrotrons, virtual cathode oscillators (vircators) and so on – see Fig. 2.11.

Vircators are capable of producing very powerful single energy pulses, which are simple in structure, small in dimensions, strong and capable of working across a relatively wide spectrum of microwave band frequencies. The fundamental idea behind a vircator lies in the acceleration of a powerful stream of electrons using a mesh anode. This powerful flow of electrons initially explodes out from the cathode (a metal cylindrical rod with a diameter of a few centimetres, Fig. 2.10.) under the influence of a high voltage

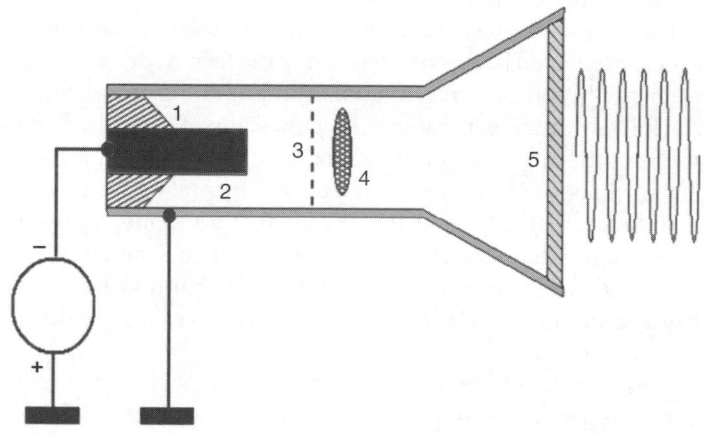

Inductive power storage
VIRCATOR

Voltage pulse (300 ns), kV	400
Current, kA	12
Power eradiation, MW	350
Duration of eradiated pulse, ns	200
Output frequency, GHz	3.1

Capacitive power storage
VIRCATOR

Voltage pulse (100 ns), kV	600
Current, kA	18
Power eradiation, MW	500
Duration of eradiated pulse, ns	80
Output frequency, GHz	3.1

Fig. 2.10 Power vircators, developed at the Tomsk Polytechnical Institute: 1 – Isolator; 2 – Metal cathode; 3 – Grid anode; 4 – Virtual cathode; 5 – Dielectric window.

pulse (hundreds of kilovolts), which lends this emission of electrons an explosive character. A significant number of electrons pass through the mesh anode, forming a space-charge cloud region beyond it. In certain conditions this space charge cloud region will oscillate around the anode.

A UHF field that is formed at the vibration frequency of the electron cloud radiates out into the open through the dielectric window. Pre-oscillation currents in the vircators, at which the generation commences, range from 1–10 kA. The vircators, see Fig. 2.11, are more suited to generating pulses with a duration of several nanoseconds in the long-wavelength part of the microwave range. Outputs ranging from 170 kW up to 40 GW in the microwave and UHF band have been obtained experimentally using this equipment. According to published data an experimental device, developing a pulse power of around 1 GW (265 kV, 3.5 kA) is capable of attacking electronic apparatus over a distance of 800–1000 m.

Even well-known devices such as the Marx high-voltage pulse generators, see Fig. 2.12, which contain a collection of high-voltage capacitors and discharge devices (80 individual modules) could be used as powerful sources of microwave radiation. In this device all the capacitors are initially charged in parallel using a high-voltage source, but as soon as the spark gaps have flashed over synchronically all the capacitors become connected in series.

Pulse currents of 6 kA at a voltage of 2.3 MV are generated in the FEBETRON-2020 mobile generator, see Fig. 2.12. Using the Marx diagram as a basis the American

Fig. 2.11 High power relativistic microwave generators based on gyrotrons, vircators and reverse wave tubes developed in different Russian Scientific Research Institutes.

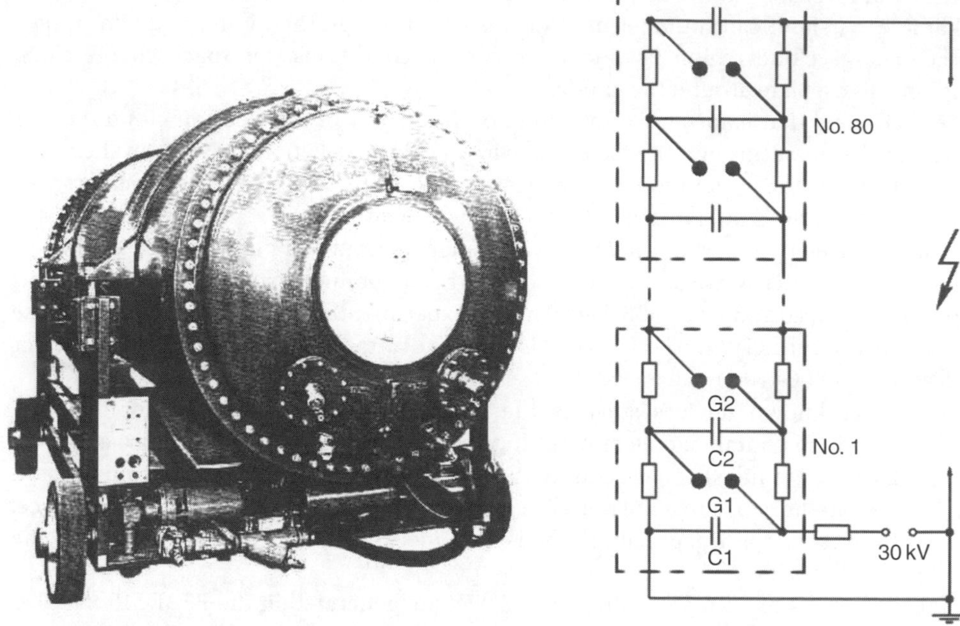

Fig. 2.12 The American FEBETRON-2020 based on a Marx generator along with a simplified diagram of the device.

company Applied Physical Electronics developed a new series of powerful ultra-compact generators that operated on a voltage of up to 1 MV with a radiated peak output of up to 6 GW, see Fig. 2.13. Fitted with a parabolic antenna, see Fig. 2.14., these devices are capable of emitting very powerful unidirectional UHF energy, which is destructive to electronics.

Another line of work in the development of IEMI is the so-called beam weapon. This weapon uses narrowly focused beams of charged or neutral particles, generated with the help of different types of accelerators based both on Earth and in satellites.

Work to create a beam weapon reached its greatest resonance soon after the announcement in 1983 by the President of the United States of America Ronald Reagan of the Strategic Defence Initiative (SDI).

The Los Alamos National Laboratory and even the Livermore National Laboratory became the centres of scientific research in this field while in Russia the Federal State Institution 'The 12th Central Scientific Research Institute of the Department of Defence of the Russian Federation' served as the centre for this research. According to statements made by several scientists, successful attempts were being made at these institutions to obtain a stream of highly energized electrons, which in terms of their output were hundreds of times more powerful than those obtained in research accelerators.

In this same laboratory, within the confines of the 'Antigon' programme, it was established experimentally that an electronic beam is distributed almost perfectly,

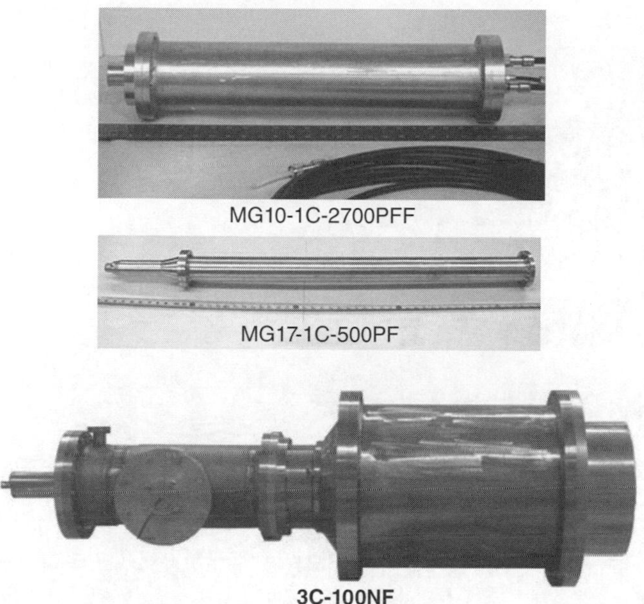

Fig. 2.13 High power compact Marx generators - the 300 kV, 1 GW MG10–1C-2700PFF), the 510 kV, 400 MW MG17–1C-500PF) and the 600 kV, 6 GW MG30–3C-100NF.

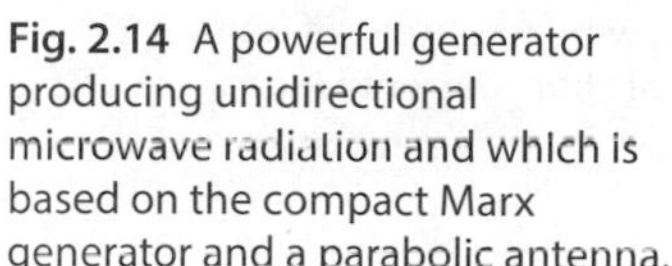

Fig. 2.14 A powerful generator producing unidirectional microwave radiation and which is based on the compact Marx generator and a parabolic antenna.

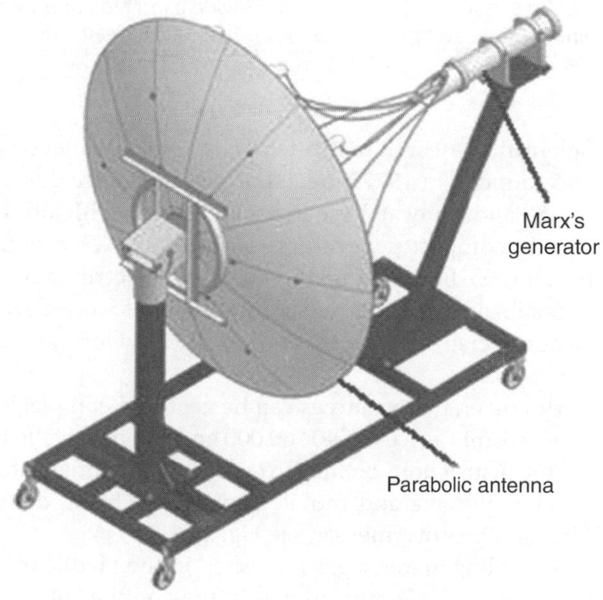

without any scattering, along an ionized channel created in advance using a laser beam in the atmosphere.

Powerful compact radiation sources that can be installed inside the cargo compartment of a lorry or even in a minibus present a serious threat. At the Tomsk

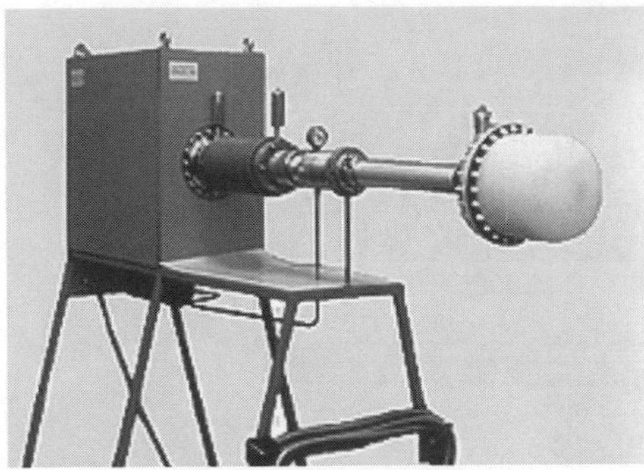

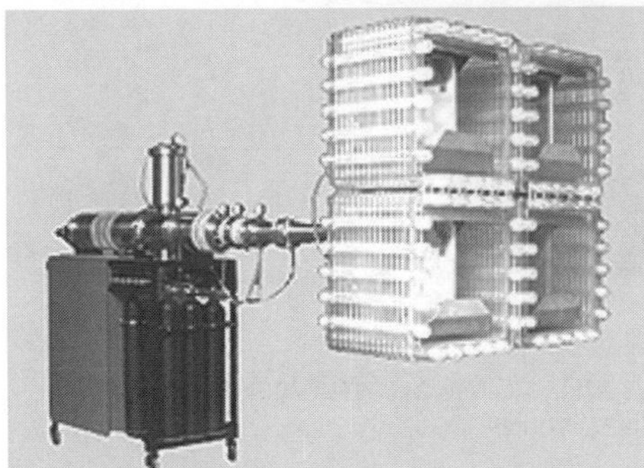

Fig. 2.15 Compact high power ultra-wide band radiation sources with an output of up to 1 GW, which were developed at the Tomsk Scientific Research Institute of High-Current Electronics of the SB RAS.

Scientific Research institute of High-voltage Electronics of the SB RAS USSR, which was founded in 1977, special procedures were developed for generating high power (giga- and terawatt) electrical pulses specifically for research, as were the sufficiently compact generators (100–1000 MW) producing linear polarized unidirectional wave beams of ultra-wide band electromagnetic radiation with a nanosecond and sub-nanosecond pulse duration that were developed under the supervision of Academician G.A. Mesyats specifically for targeting electronic equipment, see Fig. 2.15.

Moreover, these sources can be acquired completely openly at the Institute for a far from 'cosmic' price of $40–60 000 and can be installed in a minibus or even a small van. All the instructions on how to order this equipment are set out on the Institute's website. Similar portable and mobile sources are being developed and manufactured in the United States of America, see Fig. 2.16.

According to messages published in the media, in October 2012, Boeing conducted a test of a missile created within the confines of the CHAMP programme (Counter-electronics High-powered Microwave Advanced Missile Project) at the Yuta test range in almost military conditions. During the course of this testing the CHAMP missile, which was conducting a flight in accordance with a set programme, generated powerful electromagnetic pulses effectively taking electronic subsystems out of action and

Fig. 2.16 A compact source of powerful unidirectional ultra-short wave (95 GHz) radiation, developed by the American Sandia National Laboratories using technology produced by Raytheon (above) as well as powerful sources of unidirectional radiation, mounted on a Hummer jeep chassis and on a Stryker APC. It is planned to install an even more powerful system on board the AC-130 aircraft.

destroying data without inflicting any physical destruction. The CHAMP missile demonstrated that the capability existed to conduct selected, and highly calculated attacks on several targets in the course of one flight, using a High Powered Microwave. The flight was monitored from Hill Air Force Base.

CHAMP is set to become a non-lethal alternative to kinetic weapons systems and to traditional explosive weapons by striking enemy targets that are fitted with radio-electronic equipment. This missile enables an asset to be taken out of action, damaging its electronics and at the same time inflicting minimal collateral damage to the surrounding area.

According to an assessment by the head of the 'Boeing Phantom Works' programme, Kate Coleman, this technology marks a new era in contemporary warfare. In the near future this weapon could be used to take an enemy's electronic equipment and information systems out of action without recourse to air and land forces. According to a statement by

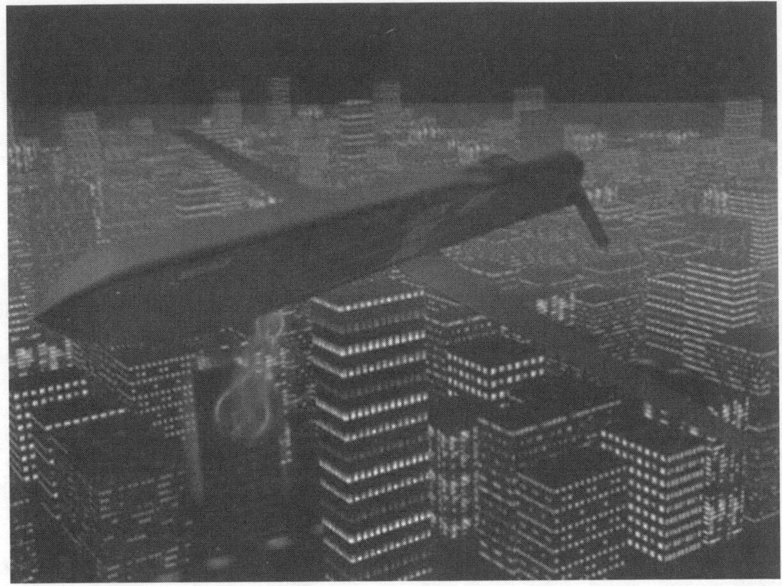

Fig. 2.17 A promotional image for a cruise missile fitted with a CHAMP electronic warhead developed by Boeing.

Boeing, this project represents directed energy technology that has been adapted by the United States Air Force laboratory for a missile platform developed by the same company, and will serve as the basis of a new family of highly effective non-lethal weapons systems.

As the principle contractor for this project Boeing is preparing the aerial platform and carrying out the final systems integration. Raytheon supplied the source of HPM microwave radiation and Sandia National Laboratories is to deliver the supply system under a separate contract with the United States Air Force's National Research Laboratory.

In a promotional film produced by Boeing about this programme, a cruise missile is shown (see Fig. 2.17) flying over a city and 'putting out' the lights. Specifically, the main control panel of a power station can be glimpsed with all the signals going out as the missile flies over.

In connection with the high technological effectiveness there are only a few companies in the world that are developing electromagnetic weapons. The global leaders are the American companies Northrop Grumman, Lockheed Martin, Raytheon, ITT and BAE Systems. In Russia the leading indigenous developer and manufacturer of radio-electronic warfare systems is the Joint Stock Company 'Radio-electronic Technology Concern.' Formed in 2009 this holding company brought together 18 enterprises under its control; scientific research institutes, design bureaux and manufacturing plants specializing in the production of air-, sea- and land-based radio-electronic warfare systems. Within the confines of the State Armaments 2011–2020 procurement programme the concern is planning to expand its presence on the radio-electronic warfare systems market.

Recently the 'Moskva-1,' 'Krasukha-2,' 'Krasukha-4' and 'Rtut-BM' systems amongst others have successfully completed state testing and have entered series production, see Fig. 2.18. These have been developed by the All-Russian Scientific Research Institute 'Gradient,' part of the concern. The 'Ranets-E' mobile microwave system (developed by

Fig. 2.18 A 1 L269 'Krasukha-2' mobile radio-electronic suppression system.

Fig. 2.19 The Ranets-E mobile high-powered UHF long-range radiation system.

the Moscow radio-technical institute) with a pulse output of 500 MW ensures that electronic assets can be hit at distances of between 12–14 km, while significant shortcomings in its operation can be observed at distances of up to 40 km, see Fig. 2.19. In actual fact, this is a powerful microwave gun specially designed to target electronic equipment. Its disadvantage lies in its considerable weight (more than 5 tonnes) and the most important consideration is the need for a pause of around 20 minutes between each 'shot'.

Information appeared recently in the press concerning a Russian analogue to the American CHAMP: the new 'Alabuga' missile system (similar missile systems were produced earlier in Russia) with a very powerful pulse emitter.

This missile's warhead, which was detonated at an altitude of 200 m, is capable of destroying all the electronics within a radius of 3.5 km. The principle difference between the Alabuga and the American CHAMP lies in its mode of operation: the Alabuga employs a single use explosively pumped flux compression generator, while the American system uses a vircator, which is capable of operating in a constant mode.

During the 1980s in the last century the well-known Soviet scientist Professor I.V. Grekhov who worked at the A.F. Ioffe Institute of Technical Physics (in St Petersburg) conducted theoretical and experimental work on the formation of high-voltage nanosecond voltage variations in standard high-voltage semiconductor diodes [2.53–2.55]. This phenomenon of the advent of overvoltages at the point at which high power semiconductor diodes switched from a conducting direction to the reverse is well-known in terms of this technology and is countered in different ways, since these overvoltages reduce the operational reliability of the diodes themselves, as well as the other elements of electronic circuits. In the work that began under the supervision of I.V. Grekhov attempts to the contrary were made to strengthen this effect and use it to generate powerful nanosecond pulses, by using semiconductor high-voltage diodes as circuit breakers in powerful pulse systems with an inductive energy storage unit.

This work was subsequently continued at the Ural branch of the Institute of Electrophysics of the Russian Academy of Sciences (in the city of Yekaterinburg). Experiments conducted in 1991–1992 by S.K. Lyubutin, S.N. Rukin and S.P. Timoshenkov on standard high voltage rectifying semiconductor diodes showed that given a specific balance in the density of direct and reverse currents and the flow time through the semi-conductor structure of a diode the timespan of the decay of the reverse current is reduced to tens of nanoseconds. Typical values for the current density in this case were tens of kiloamps per cm^2 and the flow time lies within the hundreds of nanoseconds range.

This new effect of a nanosecond interruption in ultra-high density currents in semiconductors was later named the SOS-effect (derived from Semiconductor Open Switch) [2.56]. Subsequently, a special semiconductor structure with a very strict recovery mode was developed, on the basis of which it was possible to create new high-voltage semiconductor switches, known as SOS-diodes, which had an operating voltage of hundreds of kilovolts, a switching current of tens of kiloamps, a switching time of nanoseconds and a switching frequency of kilohertz.

The typical design of an SOS-diode consists of elementary diodes assembled in series and drawn together using dielectric cotter pins between two pole plates, see Fig. 2.20 [2.57,2.58].

On the basis of SOS-diodes a series of reusable mobile compact electromagnetic pulse SOS-generators in the nanosecond range were developed at the Urals branch of the Institute of Electro-physics of the Russian Academy of Sciences that possessed record semiconductor switching parameters, see Fig. 2.21.

According to information presented on the website of the Urals branch of the Institute of Electro-physics of the Russian Academy of Sciences these generators are designed to use a solid-state energy switching system, which employs thyristors or transistors in the input device, magnetic switches in the intermediary energy compression device, and a semiconductor current breaker in the SOS-diodes in the rearmost

Fig. 2.20 A type SOS-180–12 (180 kV, 12 kA) SOS-diode.

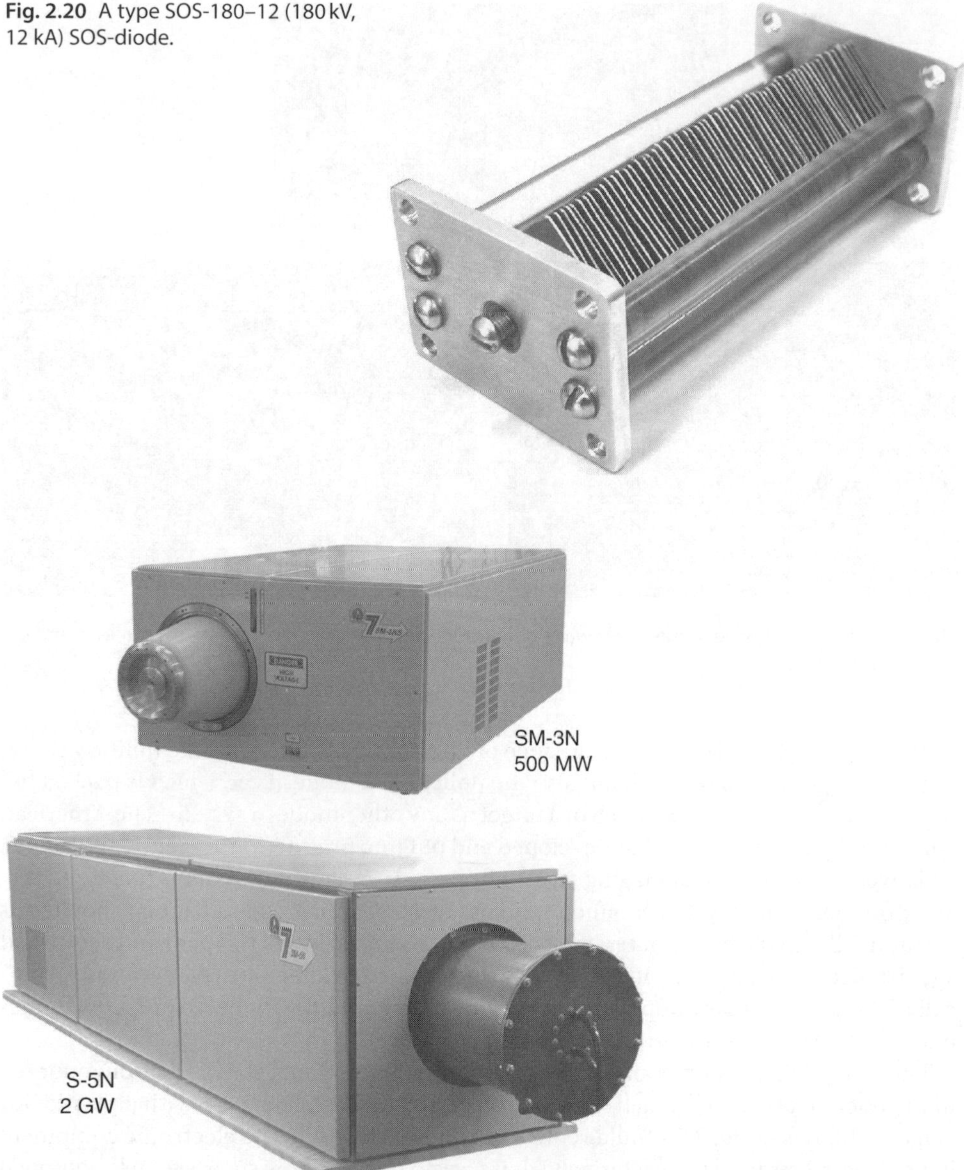

Fig. 2.21 Different types of SOS-generators developed at the Urals branch of the Institute of Electro-physics of the Russian Academy of Sciences.

power amplifier. These generators possess output parameters across the following ranges: pulse voltage amplitude 50 kV – 1 MV, pulse voltage 1–10 kA, peak output 100 MW – 4 GW, pulse duration 3–60 ns, and a pulse frequency ranging from hundreds of Hertz to individual kHz.

In certain countries (the USA and Israel among others) development work is underway on relatively low output compact electromagnetic guns, which are nevertheless

Fig. 2.22 Homemade directional microwave generators, descriptions of which are set out in popular technical journals.

capable of targeting electronics at a range of up to 100 m. Not only is the military showing an interest in these devices but also the police. The modern car, which is packed full of electronics, represents as much of a target as any other modern system. The American company Eureka Aerospace has developed and put into production an EMP car-stopper. This weapon works by damaging the microprocessor, ignition system, fuel injection, and other electronic systems found in the modern car. What would happen should this weapon fall into the hands of terrorists (sooner or later, this will happen) and they use it against electronic targets? Moreover, they do not need to look for these systems specifically. The pages of many popular technical journals abound with descriptions of homemade systems such as these, see Fig. 2.22.

These homemade sources of relatively low output directional UHF radiation mounted in the enclosed body of a small lorry or even a pickup based on a car with a plastic box van type body, see Fig. 2.23 could represent a serious threat to the electronic equipment in power systems (and not exclusively). It is easy to imagine how such a vehicle equipped with a UHF radiation source could pass a power substation with the majority of microprocessor based protection and control systems in operation in the substation, and it is hard to predict what would happen inside the power system. A single 'car' such as this could put several substations out of use in the space of a few hours, and if there are several of these vehicles?

Even industry is doing its bit for the 'common good' by producing similar devices that can be installed inside a suitcase, see Fig. 2.24, as though they have been designed especially for terrorists. It was not without reason that this was even mentioned in a report by a US congressman.

How can one forget that prophetic quote by Winston Churchill that he uttered so many years ago? 'The stone age might return on the gleaming wings of science.'

Fig. 2.23 A source of directional UHF radiation mounted on a pick-up based on a car with a plastic box van type body.

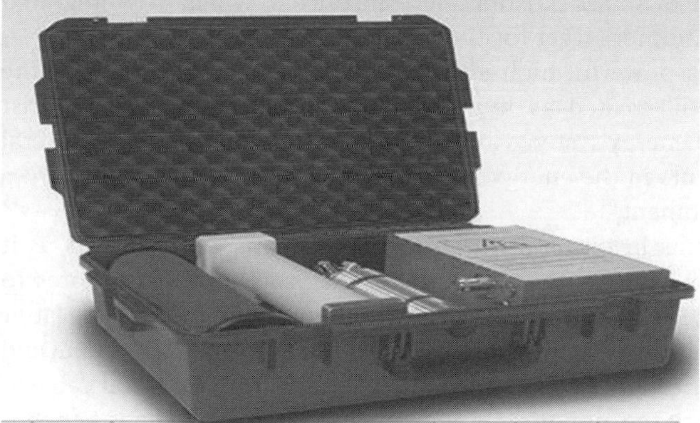

Fig. 2.24 A mobile 'electronics destruction device' known as a '2100 Series Suitcase' based on a Marx generator and produced by Applied Physical Electronics.

2.8 The Effect of HPEM on Microprocessor-based Relay Protection Systems

The various different ways in which electromagnetic radiation can penetrate electronic equipment are first and foremost via different kinds of antennae and cable entry points, electrical power supply systems, as well as currents induced inside a casing and inside wiring, and radiation passing through windows and doors made of non-conductive materials and ventilation channels. Currents, which are induced by an EMP inside surface and underground power supply cables over hundreds and thousands of kilometres can reach thousands of amps, and voltages in the open circuits in these cables can reach millions of volts. In antennae lead-ins, the length of which does not exceed a few dozen metres the currents induced by an EMP can reach several hundred amps. An EMP that passes directly through structural members made of dielectric materials (unshielded walls, windows, doors and so on) can induce currents of up to dozens of amps through internal wiring. Long-distance overhead electric power lines that absorb radiation from a wide area and deliver it directly to its destination – the input points of highly sensitive electronic equipment – represent a serious threat. The presence of transformers (both instrument and power transformers) have almost no effect on this process owing to the considerable internal capacitance between the primary and secondary windings. However, since low voltage circuits and radio-electronic apparatus normally operate at voltages of only a few volts and currents that reach several dozen milliamps, in order that they be protected reliably from EMP it is necessary to ensure a reduction in the currents and voltages at their input points by several orders of magnitude. Aside from microprocessor based relay protection systems, as strange as it may seem, optical data transfer systems, which are widely used in relay protection, possess an increased sensitivity to EMP. Specifically the controllers that transform the electrical signals into optical signals at one end of the fibre-optic communication lines (FOCLs) and restore them from optical signals at the other end of the FOCL. For example, testing using FOCUS type multiplex equipment that conformed to the IEC standards for electromagnetic compatibility showed that they cannot always withstand even standard effects without flashovers or damage. The SCADA (Supervisory control and data acquisition) system with its large number of smart sensors and transducers, which are integrated into a computer network, is another target for the effects of even weakened EMP.

If the possibility of using a powerful high-altitude nuclear explosion to attack the national energy system is still treated as hypothetical, then an attack by terrorist structures on individual local energy systems aided by a simultaneous attack on several of the most important elements in the energy system using non-nuclear EMP sources could easily happen at any moment.

The most vulnerable systems in terms of intentional electromagnetic threats as it turns out are communication systems that use broadband protocols. This relates to ATM 155, Fast Ethernet and Gigabit Ethernet, amongst others, see Fig. 2.25. The latter can be explained by an insignificant difference in the power of the legitimate signal and the interference in the upper reaches of the spectrum. The shift from a coaxial cable to a simple twisted pair cable with the aim of making the cable cheaper (which is happening everywhere today) leads to the system becoming even more vulnerable. Ethernet that employs a twisted pair cable is now being introduced in relay protection and in accordance with the strategy of Smart Grid the use of this cable in controlling all the elements in electrical power engineering is set to grow.

Discrete electronic components are much more resilient to overvoltages and other unfavourable effects, than integral microcircuits [2.60]. According to data [2.61] 75% of all the damage to microprocessor systems occurs due to the effects of overvoltages. These overvoltages with amplitudes ranging from dozens of volts to several kilovolts are a consequence of switching processes within circuits or occur as a result of electrostatic discharges, and are 'fatal' for the internal micro-elements in microcircuits and processors. According to data [2.61] standard transistors (discrete elements) can withstand a voltage from an electrostatic discharge that is almost 70 times higher than an EPROM microprocessor memory chip, for example. Computerized industrial equipment

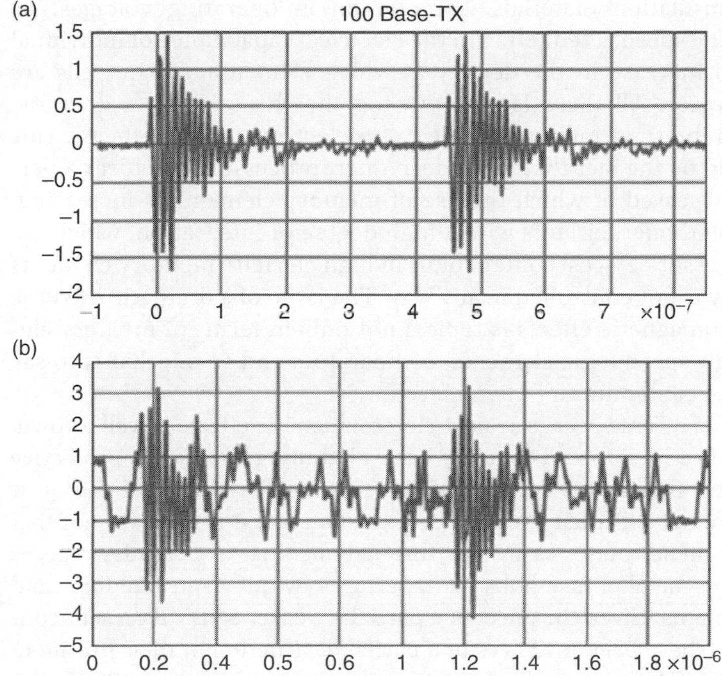

Fig. 2.25 A typical oscillogram showing induced electromagnetic interferences in an Ethernet communications link: (a) without data transfer and (b) with data transfer (the specification 100Base TX) [2.59].

(including DPR) is especially vulnerable to the effects of EMP since in essence they are built on high-density MOS devices, which are very sensitive to the effects of high voltage switching processes. An idiosyncrasy of MOS-devices is the very low level of energy (a voltage in the region of dozens of volts) that is required to damage them or to destroy them completely.

Three levels of degradation in semiconductor devices under the effects of powerful EMP have been established: a malfunction, persistent changes in the parameters, and catastrophic irreversible failures. These same irreversible failures in semiconductors occur in the main because they have overheated or as a result of a flashover [2.62–2.64]. Damage to the microprocessor or elements of the memory caused by weakened electromagnetic effects can be hidden [2.65]. This damage does not show up in any tests and can manifest itself at the most inopportune moments. Apart from that under the influence of an EMP weakened by protective measures, random reversible failures can occur, which are brought about by a spontaneous change in the contents of the memory cells. These are known as 'soft failures' or 'soft errors'. Mistakes such as these (that is to say reversible, and regenerative malfunctions) were previously unknown in electronic equipment based on discrete semiconductor elements or on standard microcircuits.

The progress made over the last few years in the field of nanotechnology has led to a significant reduction in the sizes of semiconductor elements (this refers to individual and even fractions of a micron), a reduction in the thickness of layers of semiconductor and insulation materials, a reduction in operating voltages, an increase in the operating speed, a reduction in the electrical capacitance of individual memory cells, and an increase in the density at which elementary logic cells are located in a single device. All these factors put together have led to a dramatic increase in the vulnerability of memory elements to electromagnetic effects. This problem is exacerbated by the fact that in modern microprocessor structures a persistent trend can be observed in which the use of memory elements is increasing. Many modern integrated microcircuits with a high degree of integration, which are installed in microprocessor devices, contain built in high capacity memory elements and their serviceability is not controlled in any way. The issue of a dramatic increase in sensitivity to electromagnetic effects is topical not only in terms of memory elements but also for high-speed logic elements, comparators and so on, that is to say for almost the whole of contemporary microelectronics.

The shielding action of a Faraday cage against electromagnetic effects is well-known. Buildings constructed from reinforced concrete have a lightning conductor, protective relays are housed in metal cabinets, and the DPR themselves have a metal casing. It would appear to be more of a 'Faraday Russian doll' than a cage. It is however not that simple. Firstly, high frequency pulses easily pass through openings in a Faraday cage, as well as any kind of non-metallic fuse links and apertures, windows in buildings and through ventilation systems. Given the effect of a partially weakened EMP on semiconductor equipment there have been instances of a partial destruction of their p-n-junctions, which has led to changes in their characteristics and to the advent of «flash» failures in the operation of the equipment. A considerable number of serviceable resources are linked together by these failures, and apart from that they limit confidence in the reliability of the apparatus. These 'flash' failures are at times very difficult to detect, which means it is necessary to take the equipment out of service time and time again, which leads to a significant loss of operating time, to diagnose the damage. This factor should also be taken into account when assessing the degree of protection for equipment from electromagnetic attacks, since partial or incomplete protection may lead to additional problems.

The second problem is known as the «delaying effect of EMP» and represents a very dangerous property of HPEM. This effect manifests itself in the first few minutes after the detonation of a nuclear charge or that of an electromagnetic bomb. In this space of time the EMP, having found its way through electrical systems, creates localized electromagnetic fields inside them. As these fields decay this results in rapid voltage variations, which are distributed as waves along the wires in the electrical supply system over relatively long distances from the location of the original pulse. Thirdly, external cables and wires extending for kilometres out from a protective relay cabinet and from a building all but deaden the weakening effect even of the building and the protective relay cabinets.

2.9 The Principle Technical Standards in the HEMP Field

In connection with acknowledging the overall severity of the problem of HEMP in recent years, organizations such as the International Electrotechnical Commission (IEC), CIGRE (Conference Internationale des Grands Reseaux Electriques a Haut Tension), a special commission under the US Congress, as well as European organizations have been working intensively on this problem. It is self-evident that corresponding standards and other technical documentation are required to be able to carry out productive work in this field. Some of these standards have already been drawn up by the IEC:

1) **IEC TR 61000–1-3** Electromagnetic compatibility (EMC) – Part 1–3: General – The effects of high-altitude EMP (HEMP) on civil equipment and systems.
2) **IEC 61000–1-5** High power electromagnetic (HPEM) effects on civil systems.
3) **IEC 61000–2-9** Electromagnetic compatibility (EMC) – Part 2: Environment – Section 9: Description of HEMP environment – Radiated disturbance. Basic EMC publication.
4) **IEC 61000–2-10** Electromagnetic compatibility (EMC) – Part 2–10: Environment – Description of HEMP environment – Conducted disturbance.
5) **IEC 61000–2-11** Electromagnetic compatibility (EMC) – Part 2–11: Environment – Classification of HEMP environments.
6) **IEC 61000–2-13** Electromagnetic compatibility (EMC) – Part 2–13: Environment – High-power electromagnetic (HPEM) environments – Radiated and conducted.
7) **IEC 61000–4-23** Electromagnetic compatibility (EMC) – Part 4–23: Testing and measurement techniques – Test methods for protective devices for HEMP and other radiated disturbances.
8) **IEC 61000–4-24** Electromagnetic compatibility (EMC) – Part 4: Testing and measurement techniques – Section 24: Test methods for protective devices for HEMP conducted disturbance – Basic EMC Publication.
9) **IEC 61000–4-25** Electromagnetic compatibility (EMC) – Part 4–25: Testing and measurement techniques – HEMP immunity test methods for equipment and systems.
10) **IEC 61000–4-32** Electromagnetic compatibility (EMC) – Part 4–32: Testing and measurement techniques – High-altitude electromagnetic pulse (HEMP) simulator compendium.
11) **IEC61000–4-33** Electromagnetic compatibility (EMC) – Part 4–33: Testing and measurement techniques – Measurement methods for high-power transient parameters.
12) **IEC 61000–4-35** Electromagnetic compatibility (EMC) – Part 4–35: Testing and measurement techniques – HPEM simulator compendium.
13) **IEC 61000–4-36** Electromagnetic compatibility (EMC) – Testing and measurement techniques – IEMI Immunity Test Methods for Equipment and Systems.
14) **IEC/TR 61000–5-3** Electromagnetic compatibility (EMC) – Part 5–3: Installation and mitigation guidelines – HEMP protection concepts.
15) **IEC/TS 61000–5-4** Electromagnetic compatibility (EMC) – Part 5: Installation and mitigation guidelines – Section 4: Immunity to HEMP – Specifications for protective devices against HEMP radiated disturbance. Basic EMC Publication.
16) **IEC 61000–5-5** Electromagnetic compatibility (EMC) – Part 5: Installation and mitigation guidelines – Section 5: Specification of protective devices for HEMP conducted disturbance. Basic EMC Publication.
17) **IEC 61000–5-6** Electromagnetic compatibility (EMC) – Part 5–6: Installation and mitigation guidelines – Mitigation of external EM influences.
18) **IEC 61000–5-7** Electromagnetic compatibility (EMC) – Part 5–7: Installation and mitigation guidelines – Degrees of protection provided by enclosures against electromagnetic disturbances (EM code).

19) **IEC 61000–5-8** Electromagnetic compatibility (EMC) – Part 5–8: Installation and mitigation guidelines – HEMP protection methods for the distributed infrastructure.
20) **IEC 61000–5-9** Electromagnetic compatibility (EMC) – Part 5–9: Installation and mitigation guidelines – System-level susceptibility assessments for HEMP and HPEM.
21) **IEC 61000–4-36** Electromagnetic compatibility (EMC) – Testing and measurement techniques – IEMI Immunity Test Methods for Equipment and Systems.

Some of the standards are still under development. The Institute of Electrical and Electronics Engineers (IEEE) in the United States of America also has its own standard:

IEEE P1642 Recommended Practice for Protecting Public Accessible Computer Systems from Intentional EMI.

The European Commission has also drawn up a similar document:

Topic SEC-2011.2.2–2 Protection of Critical Infrastructure (structures, platforms and networks) against Electromagnetic (High Power Microwave (HPM)) Attacks.

A separate working group has been set up around this issue at CIGRE:

CIGRE WG C4.206 Protection of the High Voltage Power Network Control Electronics against IEMI.

A great number of standards have been published by The Department of Defence of the United States of America and NATO:

1) **MIL-STD-2169B** High-Altitude Electromagnetic Pulse (HEMP) Environmental, 2012.
2) **MIL-STD-188–125–1** High-Altitude Electromagnetic Pulse (HEMP) Protection for Ground Based C41 Facilities Performing Critical. Time-Urgent Mission. Part 1 Fixed Facilities, 2005.
3) **MIL-STD-188–125–2** High –Altitude Electromagnetic Pulse (HEMP) Protection for Ground Based C41 Facilities Performing Critical. Time-Urgent Mission. Part 2 Transportable Systems, 1999.
4) **MIL-STD-461 F** Requirements for the Control of Electromagnetic Interference Characteristics of Subsystems and Equipment, 2007.
5) **MIL-STD-464C** Electromagnetic Environmental Effects. Requirements for Systems, 2010. - Test Operations Procedure Report No. 01–2-620 High-Altitude Electromagnetic Pulse (HEMP) Testing.
6) **MIL-STD-1377** Effectiveness of Cable, Connector, and Weapon Enclosure Shielding and Filters in Precluding Hazards of Electromagnetic Radiation to Ordnance (HERO), 1971.
7) **MIL-HDBK-240** Hazards of Electromagnetic Radiation to Ordnance (HERO) Test Guide, 2002.
8) **NATO AECTP-500** Ed. 4. Electromagnetic Environmental Effects Test and Verification, 2011.
9) **NATO AECTP-250** Ed.2 – Electrical and Electromagnetic Environmental Conditions, 2011.

In Russia only a few official documents concerning this issue have been published and are in the public domain:

GOST R 53111–2008 The reliable operation of the public communications network

RD 45.083–99 Recommendations for ensuring the resilience of landline telecommunications hardware to the effects of destabilizing factors.

GOST R 52863–2007 Information security. Automated systems during the conduct of a protected test of their resilience to intentional high-powered electromagnetic threats.

The first two documents relate purely to a narrow technical field, to communications systems, and are based on the requirements for everyday and not specialized standards for electromagnetic compatibility (EMC). The last document is intended for more widespread application, but is also based on the requirements of everyday standards for EMC. All the specialized requirements (which are actually those for HPEM) are marked in this standard with the letter 'X' without any indication of the technical parameters, which considering the title of the standard seems somewhat strange.

There is of course material that has not been published and is not in the public domain, such as: 'The ECC Russian Federation's standards for the resilience of apparatus, instruments, devices and equipment to the effects of IEMI and HEMP' – the decision of the State Commission for Telecommunications under the Ministry of Communications of the Russian Federation No. 143 dated 31 January 1996, as well as secret (in contrast to the West) military standards. It is true that this material does have a very limited field of application, but this is an indication in itself of the specific preoccupation among specialists with the issue of HPEM. It is however evident from the list of standards set out above that the standards for the resilience of apparatus to the effects of HPEM and the methods for testing this have long since stopped being a secret and have been published openly in the West along with both the general IEC standards and the military ones (with the rare exception of MIL-STD-2169B for example). Russia's policy concerning a purely technical issue, as a result of which the majority of Russian specialists in many aspects of civilian technology (such as the electrical power engineering field for example) and who are linked in one way or another to the problems of HPEM but do not have the first clue about it, could hardly be called logical and plausible.

In recent years a whole series of dissertations dedicated to this issue have been written in Russia [2.66–2.69] although they all relate to the effect of HPEM on communications systems this does not change the nature of the problem.

References

2.1 Gurevich V.I. The vulnerability of DPR. Problems and solutions. - M.: *Infra-Inzheneriya*, 2014–256 pp.

2.2 Operation Dominic, Fish Bowl Series, Debris Expansion Experiment. Air Force Weapons Laboratory. Project Officer's Report, Project 6.7, Report AD-A995 428, POR-2026 (WT-2026), 10 December 1965.

2.3 Loborev V.M., Up to date state of the HEMP problems and topical research directions, in *Euro Electromagnetic Conf. (EUROEM), Bordeaux, France*, June 1994, pp. 15–21.

2.4 Kompaneets A.S., Radio emission from an atomic explosion - *Soviet Physics JETP*, December 1958.

2.5 Karzas W.J., Latter R., Electromagnetic radiation from a nuclear explosion in space - *Physical Review*, Vol. 126 (6), pp. 1919–1926, 1962.

2.6 Karzas W.J., Latter R., EMP from high-altitude nuclear explosions, Report No. RM-4194, Rand Corporation, March 1965.

2.7 Karzas W.J., Latter R., Detection of electromagnetic radiation from nuclear explosions in space - *Physical Review*, Vol. 137, March 1965.

2.8 Inston H.H., Diddons R.A. *Electromagnetic Pulse Research. - ITT Research Institute Project T1029*, Chicago, Illinois 60616, Final Report, September 1965.

2.9 *DASA EMP (electronic pulse) Handbook, by United States Defense Atomic Support Agency*. Information and Analysis Center, National Government Publication, Santa Barbara, CA, 1968.

2.10 Electromagnetic Pulse Problems in Civilian Power and Communications, Summary of a seminar held at Oak Ridge National Laboratory, August 1969, sponsored by the U.S. Atomic Energy Commission and the Department of Defense, Office of Civil Defense.

2.11 EMP Threat and Protective Measures. – Office of Civil Defense, TR-61, August, 1970.

2.12 Parks G.S., Dayaharsh T.I., Whitson A.L., A Survey of EMP Effects During Operation Fishbowl, Defense Atomic Support Agency (DASA), Report DASA-2415, 1970.

2.13 Nelson D.B., *A Program to Counter the Effects of Nuclear Electromagnetic Pulse in Commercial Power Systems*, Oak Ridge National Laboratory, Report ORNL-TM-3552, Part 1. 8, October 1972.

2.14 Marable J.H., Baird J.K., Nelson D.B., Effects of Electromagnetic Pulse of a Power System, Oak Ridge National Laboratory, Report ORNL-4836, December 1972.

2.15 Sandia Laboratories, 'Electromagnetic Pulse Handbook for Missiles and Aircraft in Flight', SC-M-71 0346, AFWL TR 73–68, EMP Interaction Note 1–1, September, 1972.

2.16 Rickets L.W., *Fundamentals of Nuclear Hardening of Electronic Equipment -* John Wiley & Sons, Inc., 1972.

2.17 Baird J.K., Frigo N.J., Effects of Electromagnetic Pulse (EMP) on the Supervisory Control Equipment of a Power System, Oak Ridge National Laboratory, Report ORNL-4899, October 1973.

2.18 Rickets L.W., Bridges J.E., Miletta J, *EMP Radiation and Protective Techniques*, John Wiley & Sons, New York, 1976.

2.19 United States High-Altitude Test Experiences: A Review Emphasizing the Impact on the Environment, Report LA-6405, Los Alamos Scientific Laboratory. October 1976.

2.20 Glasstone S., Dolan P.J., *The Effects of Nuclear Weapons*. U.S. Department of Defense, Washington, DC, 1977.

2.21 Longmire C.L., On the electromagnetic pulse produced by nuclear explosions - *IEEE Trans. on Electromagnetic Compatibility*, Vol. EMC-20, No. 1, pp. 3–13, February 1978.

2.22 Sollfrey W., *Analytic Theory of the Effects of Atmospheric Scattering on the Current and Ionization Produced by the Compton Electrons from High Altitude Nuclear Explosions*, Rand Corp., R-1973-AF, 1977.

2.23 Butler C., *EMP Penetration Handbook for Apertures, Cable Shields, Connectors, Skin Panels, AFWL-TR-77–149*, Air Force Weapons Laboratory (The Dikewood Corporation), December 1977.

2.24 HEMP Emergency Planning and Operating Procedures for Electric Power Systems, Oak Ridge National Laboratory, Report ORNL/Sub/91-SG105/1, 1991.

2.25 Impacts of a Nominal Nuclear Electromagnetic Pulse on Electric Power Systems, Oak Ridge National Laboratory, Report ORNL/Sub/83–43374, 1991.

2.26 HEMP-Induced Transients in Electric Power Substations. Oak Ridge National Laboratory, Report ORNL/Sub-88-SC863, February 1992.

2.27 Report of the Commission to Assess the Threat to the United States from Electromagnetic Pulse (EMP) Attack. Critical National Infrastructures, April 2008.

2.28 High Altitude Electromagnetic Pulse (HEMP) and High Power Microwave (HPM) Devices: Threat Assessments. CRS Report for Congress, July 2008.

2.29 The Early-Time (E1) High-Altitude Electromagnetic Pulse (HEMP) and Its Impact on the U.S. Power Grid, Report Meta-R-320, Metatech Corp., January 2010.

2.30 The Late-Time (E3) High-Altitude. Electromagnetic Pulse (HEMP) and Its Impact on the U.S. Power Grid, Report Meta-R-321, Metatech Corp., January 2010.

2.31 Intentional Electromagnetic. Interference (IEMI) and Its Impact on the U.S. Power Grid, Report Meta-R-323, Metatech Corp., January 2010.

2.32 High-Frequency Protection Concepts for the Electric Power Grid, Report Meta-R-324, Metatech Corp., January 2010.

2.33 Protection of High Voltage Power Network Control Electronics Against Intentional Electromagnetic Interference (IEMI), Report CIGRE Working Group C4.206, November 2014.

2.34 IEC TR 61000–1-3 Electromagnetic compatibility (EMC) – Part 1–3: General – The effects of high-altitude EMP (HEMP) on civil equipment and systems.

2.35 IEC 61000–2-9 Electromagnetic compatibility (EMC) - Part 2: Environment - Section 9: Description of HEMP environment - Radiated disturbance. Basic EMC publication.

2.36 IEC 61000–2-10 Electromagnetic compatibility (EMC) - Part 2–10: Environment - Description of HEMP environment - Conducted disturbance.

2.37 IEC 61000–2-11 Electromagnetic compatibility (EMC) - Part 2–11: Environment - Classification of HEMP environments.

2.38 IEC 61000–2-13 Electromagnetic compatibility (EMC) - Part 2–13: Environment - High-power electromagnetic (HPEM) environments - Radiated and conducted.

2.39 IEC/TR 61000–5-3 Electromagnetic compatibility (EMC) - Part 5–3: Installation and mitigation guidelines - HEMP protection concepts.

2.40 IEC/TS 61000–5-4 Electromagnetic compatibility (EMC) - Part 5: Installation and mitigation guidelines - Section 4: Immunity to HEMP - Specifications for protective devices against HEMP radiated disturbance. Basic EMC Publication.

2.41 IEC 61000–5-5 Electromagnetic compatibility (EMC) - Part 5: Installation and mitigation guidelines - Section 5: Specification of protective devices for HEMP conducted disturbance. Basic EMC Publication.

2.42 IEEE P1642 Recommended Practice for Protecting Public Accessible Computer Systems from Intentional EMI.

2.43 Topic SEC-2011.2.2–2 Protection of Critical Infrastructure (structures, platforms and networks) against Electromagnetic (High Power Microwave (HPM)) Attacks, European Commission Security Research Program, 2010.

2.44 MIL-STD-188–125–1. High-Altitude Electromagnetic Pulse (HEMP) Protection for Ground-Based C^4I Facilities Performing Critical Time-Urgent Missions, Department of Defense, 1994.

2.45 MIL-STD-461E. Requirements for the Control of Electromagnetic Interference Characteristics of Subsystems and Equipment, Department of Defense, 1993.

2.46 MIL-STD-464C. Electromagnetic Environmental Effects Requirements for Systems, Department of Defense, 1997.

2.47 MIL-STD-2169B. High Altitude Electromagnetic Pulse (HEMP) Environment, Department of Defense, 1993.

2.48 MIL-Hdbk-423. Military Handbook: High Altitude Electromagnetic Pulse (HEMP) Protection for Fixed and Transportable Ground-Based C41 Facilities, Vol. 1: Fixed Facilities Department of Defense, 1993.

2.49 High Altitude Electromagnetic Pulse (HEMP) Testing, Test Operations Procedure 01–2-620, U.S. Army Test and Evaluation Command, 2011.

2.50 Gurevich V.I. Enhancing the resilience of energy systems to HEMP - a vital task for our age - *Energoexpert*, 2015.

2.51 Belous V. The threat of the deployment of EMP weapons for military and terrorist purposes - *Nuclear Control*, 2005 No. 1(75), Vol. 11, pp. 133–140.

2.52 Burlakov A.A., Kirbadin V.V. On the influence of solar storms on the reliability of energy systems - abstracts from the *International Scientific-technical Congress 'Power Engineering in a Global World,' Krasnoyarsk*, 16–18 June 2010, pp. 32–33.

2.53 Grekhov I.V., Yefanov V.M., Kardo-Sysoyev A.F., Shenderey S.V. The formation of nanosecond voltage variations in semiconductor diodes with a drift recovery system - *Letters to the TPJ*, 1983, Vol. 9 (7). pp. 435–439.

2.54 Tuchkevich V.M., Grekhov I.V. New principles in high power commutation using semiconductor devices. L: *Science*, 1988, p. 117.

2.55 Grekhov, I.V. High power pulse commutation using semiconductor devices, in: *The Physics and Technology of Powerful Pulse Systems*, E.P. Velikhova (ed.). M.: Energoatomizdat, 1987, p. 237.

2.56 Slyusar V. Ultra-powerful electromagnetic pulse generators in information wars - *Electronics: Science, Technology, Business*, 2002, Issue 5, pp. 60–67.

2.57 Slovikovskiy B.G. *Small-scale High-voltage Nanosecond Pulse Generators Based on SOS-diodes*. Synopsis of a Thesis for the Degree of Candidate of Technical Sciences, Yekaterinburg, 2004.

2.58 Gurevich V.I. Optoelectronic transformers: A panacea or a private solution to private problems - *The Electro-Technical News*, 2010, No. 2, pp. 24–28.

2.59 Kirichek R.V. *Research into the Influence of Very Fast Acting Pulses on the Data Transfer Process in Ethernet networks* - Synopsis of a Thesis for the Degree of Candidate of Technical Sciences, 5 December 2013, St. Petersburg, 2011.

2.60 Gurevich V.I. The reliability of DPR: Myths and reality - *Problems of Power Engineering*, 2008, No. 56, C.47–62.

2.61 Clark O.M., Gavender R.E. Lighting protection for microprocessor-based electronic systems - *IEEE Transactions on Industry Applications*, Vol. 26, No. 5, 1990.

2.62 Bludov S.B., Gladetskiy N.P., Kravtsov K.A. et al. The generation of very fast acting powerful uhf pulses and their effects on electronic technology - *The Physics of Plasma*, 1994, Vol. 20 (7, 8), pp. 712–717.

2.63 Panov V.V., Sarkisyan A.P. Certain aspects of the problem of creating functional UHF weapons - *Foreign Radio-Electronics*, 1993, 10, 11, 12, pp. 3–10.

2.64 Antipin V.V., Godovitsyn V.A., Gromov D.V., Kozhevnikov A.S., Ravayev, A.A. The influence of powerful microwave pulse interference on semiconductor devices and integrated microcircuits - *Foreign Radio-Electronics*, 1995, 1, pp 37–53.

2.65 Phadke A.G. Hidden Failures in electric power systems - *International Journal of Physical Infrastructures*, Vol. 1, No. 1, 2004.

2.66 Voskobovich A.A. Methods of ensuring the resilience of future radio-relay, tropospheric, and satellite communication systems to the effects of powerful electromagnetic pulse interference. 5 December 2013 - Moscow, 2002.

2.67 Voskobovich A.A. Developing methods of assessing the resilience of telecommunications systems to the effects of ultrabroadband electromagnetic pulses. 5 December 2013 - Moscow, 2003.

2.68 Akbashev, B.B. Theoretical and experimental methods of assessing the resilience of terminals to the effects of ultrabroadband electromagnetic pulses. 5 December 2013 - Moscow, 2005.

2.69 Yakushin S.P. Methods and techniques for assessing the effect of high energy electromagnetic pulses on telecommunications networks. 5 December 2013 - Moscow, 2004.

3

Methods and Techniques of Protecting DPR from EMP

3.1 The Sensitivity of DPR to Electromagnetic Threats

The problem of the electromagnetic compatibility (EMC) of electronic apparatus has developed along with the equipment itself, in as much as some of the assemblies perform in such a way that they act as receptors for electromagnetic emissions, while others act as sources of radiation. These problems have arisen as a result both of the interactive influence of some assemblies on others inside the equipment, and the influence of different types of external radiation on electronic apparatus. For decades the problem of EMC has been the prerogative of specialists in electronics, radio-electronics and communications. All of a sudden, in the course of the last 15–20 years this issue has also become very topical in the field of electrical power engineering. Naturally, relatively large electromagnetic fields have always existed around electrical power equipment. However, the electro-mechanical automation, control and relay protection equipment that has been in use for decades was barely susceptible to these fields and therefore no serious problems with EMC arose. The last two decades have been characterized by an intensive shift from electro-mechanical to digital protective relays (DPR) as well as automation in electrical power engineering. Moreover, this transfer process is not only being conducted as new substations and power stations are being constructed but by replacing old electro-mechanical protective relays in old substations that were built at a time when nobody could even have envisaged that microprocessor based technology in the shape of ultramodern DPR would be used in them. The latter have turned out to be highly sensitive to electromagnetic interference 'in the air', which affects operational current circuits, voltage circuits and transformer circuits. There have been cases in which a DPR has failed simply because of a mobile phone [3.1]. Other typical examples of failures in DPR have occurred in the operational Mosenergo facilities at the Ochakovskiy and Zubovskiy substations. The operating algorithm for the protection equipment has been destroyed by lightning, an excavator working nearby, as well as by arc welding and several other sources of interference. When the Lipetsk substation was being brought on stream, where some $1 500 000 had been spent on acquiring DPR, problems with the microprocessor equipment meant that this energy site could not be brought into operation for 6 months. In the end the substation was put into operation using a traditional protection system [3.2]. In practice it was necessary to deal with instances in which, for example, short circuits of around 110 kV caused failures in the operation of the protection systems of around 330 kV, and

Protection of Substation Critical Equipment Against Intentional Electromagnetic Threats,
First Edition. Vladimir Gurevich.

the voltage types would penetrate (via the common operational current circuits) the input points of the relay protection apparatus, that operated on a different voltage type one after another [3.3].

Malfunctions in relay protection due to insufficient EMC, according to Mosenergo data, account for up to 10% of all failures and principally affect relays that use a microelectronic or microprocessor element base [3.4]. Such a high percentage of instances of failures due to insufficient EMC arose as a result of the fact that the sensitivity of DPR to electromagnetic interference is much higher than traditional electro-mechanical protection systems. For example, according to data [3.4], if energy of 10^{-3} J is required for an electro-mechanical relay to fail then just 10^{-7} J is required for integral microcircuits to fail. The difference is to the order of four or 10 000 times more.

The degree of damage is dependent on the resilience both of the individual elements of the circuit and the energy in high power interference as a whole, which could be absorbed by a circuit without causing any defects or failures. Although switching interference from an inductive loading with an amplitude of 500 V represents a more than twofold overvoltage for an electromagnetic relay with a 230 V alternating current coil for example, it would be unlikely to lead to the failure of a relay owing to the strength of the resilience of electro-mechanics to interference of this nature and is a consequence of the short duration of this interference (which lasts for microseconds). However, the situation with respect to a microcircuit fed by a 5 V direct current source is different. Pulse interference with an amplitude of 500 V exceeds the feed voltage for this electronic component by more than 100 times and leads to the unavoidable failure and subsequent destruction of the equipment. The resilience of microcircuits to overvoltages is much less than that of an electro-magnetic relay [3.5]. Pulse overvoltages that occur during lightning strikes and during switching in power equipment are capable of damaging and destroying both electronic equipment and entire systems. Statistics gathered over a number of years confirm that instances of this damage double every 3–4 years [3.5]. This statistic fits in well with the so-called Moore's Law [3.6], which proved back in 1965 that the number of semiconductor components in microchips doubles approximately every two years and this trend has been maintained over many years. If some 20 years ago so-called TTL (Transistor-Transistor Logic) microcircuits contained 10–20 elements per mm^2 and had a typical feed voltage of 5 V today, popular microcircuits can contain almost 100 CMOS (Complementary Metal-Oxide Semiconductor) transistors per mm^2 of surface area, and have a feed voltage of only 1.2 V. The newest solid-state technology, such as SOS (Silicon-On-Sapphire), for example, raises the density of elements up to 500 per mm^2 of surface area [3.7]. Evidently, an even smaller feed voltage is needed for these microcircuits. Moreover, it is very evident that as the degree of integration in microelectronics increases the resilience of its components to high-voltage pulse overvoltages decreases due to a reduction in the thickness of the insulating layers and a reduction in the operating voltages of semiconductor elements.

Since low energy interference often gives rise to high energy interference the most common reaction of a DPR to the effects of electromagnetic interference would not be the destruction of the equipment, but a malfunction or a short failure with the subsequent restoration of the defective function, see Fig. 3.1.

This means that a DPR that has failed in a substation would perform perfectly under examination in a laboratory and it would be impossible to establish the reason behind its failure. A statistic gathered by the representatives of Japan's largest DPR manufacturers demonstrates this tendency of these relays sharply, see Fig. 3.2. [3.8].

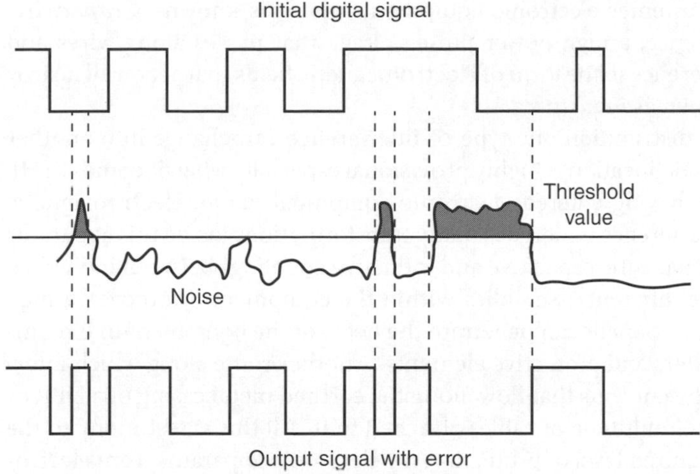

Fig. 3.1 The effect of low power interference on the operation of a digital device.

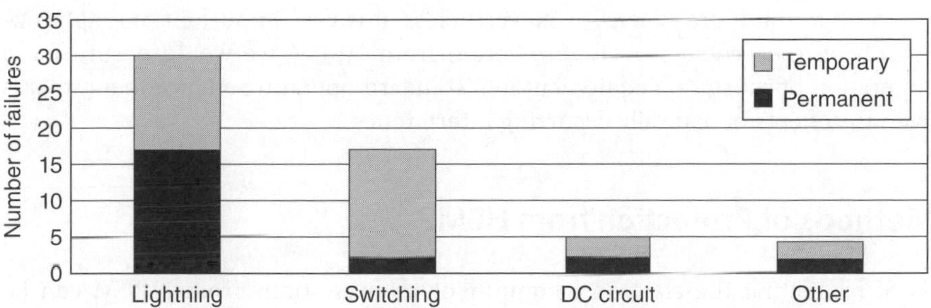

Fig. 3.2 Data provided by Japanese manufacturers on the damageability of DPR as a result of electromagnetic interference.

As is evident from Fig. 3.2, short one-off malfunctions (failures) in DPR are prevalent in most cases. According to their data malfunctions of this nature account for almost 70% of the total instances of damage to DPR and, what is more, up to 80% of these failures occur in integrated microcircuits.

According to evidence [3.4] provided by the Open Joint Stock Company, Mosenergo, there is sufficient confirmation in practice of the negative influence of electromagnetic interference on the operation of DPR. This is demonstrated most dramatically in the integration of SIEMENS relay protection equipment into TPP-12 of Mosenergo, in accordance with a project conducted by the Atomenergoproekt institute. No requirement for EMC was taken into consideration during the design stage. As a result of interference more than 400 false data signals on the discrete and analogue DPR inputs were recorded during the period from August to December 1999 alone [3.4]. Furthermore, it is worth bearing in mind that the cost of each failure in a DPR is 10 times higher than the cost of the failure of a single electromechanical relay owing to the concentration of a large number of functions in each relay.

Interference, by which short (that is to say HF) pulse overvoltages from high-voltage OPLs enter low-voltage circuits via capacitive couplings between the cables and between transformer windings can enter electronic equipment inputs, is known as *capacitive*. Aside from this interference, a high power pulse voltage that passes along wires and cables also creates interference in the form of electromagnetic fields that act on all nearby cables. This effect is known as *inductive*.

In the course of their distribution one type of interference can change into another repeatedly, therefore this delineation is highly provisional especially when it comes to HF processes. Furthermore, having entered electronic equipment via an electromagnetic field or via cables the interference undergoes many transformations inside this apparatus owing to the presence of parasitic capacitive and inductive couplings between individual elements or between the different assemblies within the equipment. Moreover, a high frequency interference component can penetrate the heart of the equipment in circumvention of established filters and protective elements. Another route along which interference can enter is via the currents that flow along the earthed metal casing of a DPR or the earthed shields of the multitude of cables attached to it. All this is testimony to the fact that ensuring the required level of protection for electronic apparatus from electromagnetic interference is extremely complicated even when it comes to naturally occurring, and not artificial interference. When it comes to protecting electronic equipment from HPEM, and particularly from HEMP, this problem becomes even more complicated and the solution more expensive. Nevertheless, it is very important to emphasize that by resolving the problem of the protection from HPEM we would ensure more reliable operation of substation equipment in a standard operating environment, that is to say given the effects of naturally occurring interference.

3.2 Methods of Protection from HEMP

Methods of protecting the electronic equipment in substations from HPEM can be divided into passive, active and administrative and technical methods.

Passive protection methods assume the use of additional external protection measures (technology, materials, elements etc.), which are not directly linked to the algorithms or the operating mode of the equipment that is being protected. They include special cabinets and specially shielded cables, overvoltage suppressors and filters, enhanced earthing systems, special construction materials, conductive paints, varnishes, film coatings, metallized mats and shutters and finally, the use of special elements and materials in the design of the electronic equipment itself.

Active protection methods assume the use of external devices, and algorithms, the operation of which is directly linked to the algorithm and to the operating mode of the equipment being protected. To use the example of DPR it can be said that specially developed simplified protective relay starting elements that have been manufactured on electromechanical principles, and which possess enhanced resilience to HPEM could be included in these protection methods. These starting elements do not react to short EMPs and cannot be damaged by them. They do however react to one of the emergency parameters (current, and voltage), which put the DPR into operation, ensuring the realization of the required characteristics for the equipment as a whole.

Administrative and technical methods include the administrative and technical facilities for the specific storage of spare parts for electronic apparatus, ensuring that damaged equipment can be repaired and restored as soon as possible, standardization of the design and the software used in electronic equipment, specifically DPR, thus ensuring the interchangeability of internal blocks and modules, as well as universalization of the methods for testing the serviceability of electronic equipment using special testing systems, thereby reducing the time taken to restore the equipment to operational condition.

In selecting protection systems to protect against EMP and when using them in practice it is advisable to bear in mind that one form of protection alone is not sufficient to ensure effective protection. Only combined and integrated use of different methods and protection systems can ensure the reliable protection of electronic equipment. Thus the use of a cabinet that is specially protected from EMP would not, for example, protect the sensitive equipment from EMP penetrating the wiring that runs into the cabinet itself and which is connected to the inputs into this equipment. The same goes for the use of special filters that protect against EMP entering wiring that would not provide protection from electromagnetic emissions entering the equipment through a window or through incisions in the casing. The use of protected cabinets and filters, however, does not solve the issues surrounding the threat to the equipment from EMP entering through the earthing system.

References

3.1 Shalin A.I. The effectiveness of the new relay protection device - *Russian Energy and Industry*, 2006, No. 1 (65).

3.2 Prokhorov A. Intelligence is our main competitive advantage (interview with General Director of OJSC ChEAZ M.A. Shurdov) - *The Equipment, Market, Supply, Price*, 2003, No. 4.

3.3 Kuznetsov M., Kungurov D., Matveev M., Tarasov V. Input circuits in relay protection devices. Issues surrounding high surge overvoltage protection - *Electro-Technical News*, 2006, No. 6 (42).

3.4 Borisov R. Negligence of EMC may be disastrous - *Electrical Engineering*, 2001, No. 6 (12).

3.5 Pravosudov P. Trabtech – Surge overvoltage protection technology - *Components and Technologies*, 2003, No. 6.

3.6 Moore G.E. Cramming more components onto integrated circuits - *Electronics*, 1965, Vol. 38, No. 8.

3.7 Nailen R.L. How to combat power line pollution – *Electrical Apparatus*, December 1984.

3.8 Matsumoto T., Kurosawa Y. Usui M., Yamashita K., Tanaka T. Experience of numerical protective relays operating in an environment with high-frequency switching surge in Japan - *IEEE Transactions On Power Delivery*, 2006, Vol. 21, No. 1.

4

Passive Methods and Techniques of Protecting DPR from EMP

4.1 Cabinets

The ideal protection from EMP would be to fully insulate electronic equipment from the outside world and to cover the entire facility in which it is housed with a solid, thick walled, ferromagnetic screen. It is clear, however, that it would be impossible to realize this protection for DPR on a practical level.

Therefore, in practice more reliable protection methods need to be used such as current bearing networks or current bearing film coatings for the windows, cellular metallic designs for the air intake and ventilation openings and special electrically conductive lubricants and linings made from electrically conductive rubbers, applied around the perimeters of the doors and hatches.

Today specially designed metal cabinets (Fig. 4.1), which are able to considerably weaken electromagnetic emissions, are well represented in the marketplace. Standard cabinets made from ordinary laminated steel and which do not contain windows or slits, are able to weaken EMP considerably. However, the use of zinc-coated mounting panels to manufacture the cabinets and the special electrically conductive sealants and linings significantly enhance the efficiency of these cabinets in as much as the zinc coating enables the voltages to be equalized over a wide area (the specific resistance of steel is 0.103–0.204 $\Omega \times mm^2/m$ while the specific resistance of zinc is 0.053–0.062 $\Omega \times mm^2/m$). Aluminium has an even lower resistance (0.028 $\Omega \times mm^2/m$). Therefore, many companies manufacture their monoblock cabinets using a special alloy called ALUZINC 150 (Aluzinc® is the registered trade mark used by the Arcelor concern) – this is a steel that has a coating that consists of 55% aluminium, 43.4% zinc and 1.6% silicon). The surface of the cabinet that contains this coating ensures a high degree of reflection of electromagnetic emissions. Cabinets made from this material are manufactured and exported to many countries around the world by Sarel (today this company is known as Schneider Electric Ltd, UK). Similar cabinets designed to provide protection from EMP are today produced by other companies: Canovate Group, R.F. Installations, Inc.; Universal Shielding Corp; Eldon; Equipto Electronics Corp.; ATOS; MFB; European EMC Products Ltd; Amco Engineering; Addison and many others. These sorts of cabinets typically weaken radiation by 80–90 dB at a frequency range of 100 KHz – 1 GHz.

The problem is that in reality relay protection cabinets unavoidably have a number of openings to allow for the entry and exit of dozens of cables. This inevitably leads to

Protection of Substation Critical Equipment Against Intentional Electromagnetic Threats, First Edition. Vladimir Gurevich.

Fig. 4.1 A control cabinet with enhanced protection from EMP fitted with special mesh, lined with an electrically conductive rubber gasket, special interfacing and linking elements, shielded ventilation windows and so on (Equipto Electronics Corp.).

a reduction in the efficiency of the screening. Apart from that DPR located inside a cabinet, as well as the cabinets themselves should be earthed. However, the existing earthing system at substations is far from sufficient in its reaction to HEMP and despite expectations is not conducive to providing protection for electronic apparatus from this threat.

4.2 The Earthing of Sensitive Electronic Apparatus

> *'Earthing is the worst understood aspect of automation...*
> *Today the solution to the problem of earthing stands*
> *on a boundary between understanding, intuition, and luck.'*
>
> Viktor Denisenko, Doctor of Technical Sciences and Chief Designer of the Scientific Technical Laboratory of Design Automation

The modern earthing system for DPR, which it is thought was designed correctly, employs a multipoint layout and uses an equipotential surface, see Fig. 4.2. The metal structural elements within the cabinets themselves can be used as an equipotential surface, see Fig. 4.3.

Unfortunately, in electrical power engineering assets (such as substations and power stations) that cover a wide area even these earthing methods are not sufficiently effective in view of the need to earth a number of different electrical installations that are situated a considerable distance from one another, and which are at different points on

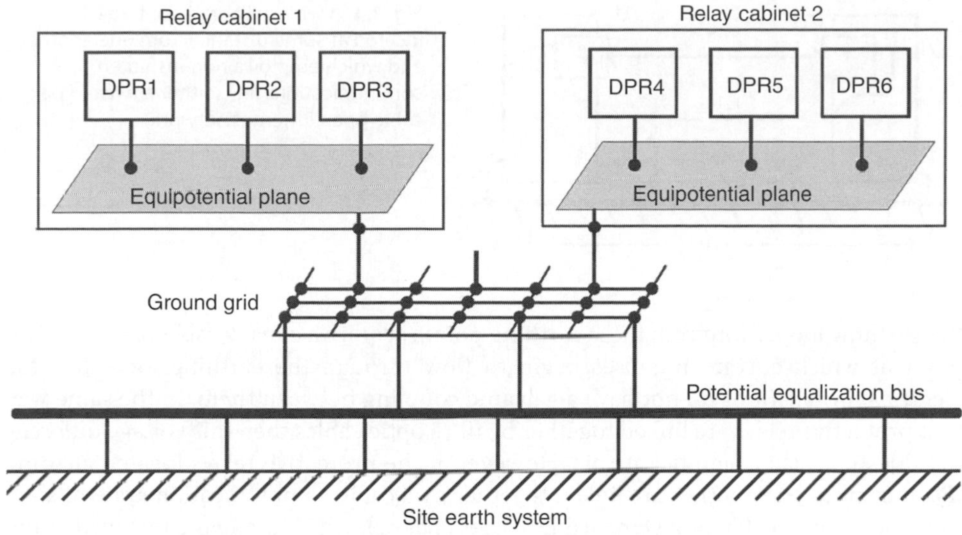

Fig. 4.2 A diagram of multi point earthing of DPR using an equipotential surface.

Fig. 4.3 A DPR earthing unit installed in metal cabinets that make use of an equipotential surface. 1 – DPR in metal casings; 2 – copper earthing ties; 3 – A structural element within the metal cabinet that performs the role of an equipotential surface.

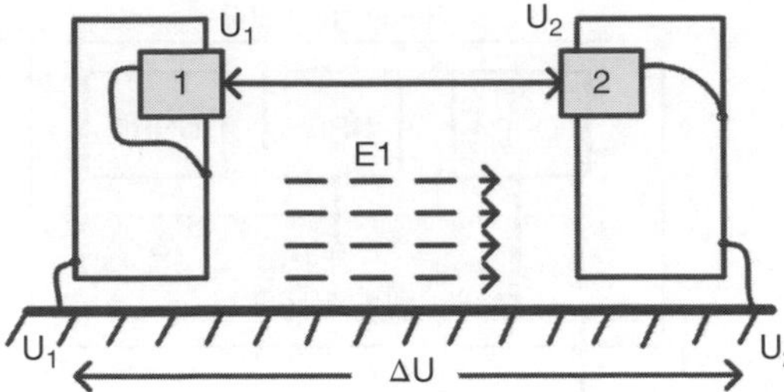

Fig. 4.4 A wiring diagram for two DPRs located at some distance from one another and which employ a non-insulated communications link (either a twisted pair cable or an Ethernet network).

the earthing loop. Moreover, these earthing points acquire a considerable voltage at the point at which current impulses begin to flow through the earthing loop. If these electrical installations do not have a galvanic coupling between them, in the same way that protective relays are linked together by fibre optic cables then this voltage differential does not play a significant role. However, if the protective relays located at some distance from one another are linked together by a cable communication system (such as a twisted pair cable or a standard Ethernet channel, which are being installed at the moment with the aim of making the supply of electricity cheaper) then high voltages can end up being applied to the low voltage assemblies within this system, and this inevitably leads to the system being damaged, that is to say in many cases to the failure of the protective relay, see Fig. 4.4.

According to information [4.1]: 'The bigger the area that an asset being protected occupies, then the greater the potential for problems.'

It is well known that there are two types of earthing: so-called 'functional earthing' (or working earthing) and 'protective earthing'. It follows even from the names for these types of earthing that the first of these is designed merely to ensure that the equipment functions normally while the second is designed exclusively to ensure the electrical safety of personnel. In reference [4.2] it is affirmed that functional earthing is necessary to ensure the serviceability of DPR and that different methods of earthing and of testing this equipment are under discussion. In reality, several DPR printed circuit boards have printed conductor elements that have been cleaned and coated with silver and have also been made wider, so that when the board is installed into the casing slot they come into contact with special springs, ensuring that these printed conductors are in contact with the earthed DPR casing, see Fig. 4.5.

However, is functional earthing really necessary for a DPR to function normally, as well as all the input and output circuits that have been well insulated from the ground and from other electrical installations (when they have been fitted with fibre optic cables to link the terminals together)? The serviceability of the internal DPR circuits is not linked in any way to the presence or otherwise of earthing. As far as the effectiveness of the protection of sensitive electronic circuits in DPR from the threat from external electromagnetic fields using a metal casing that is designed to act as a so-called 'Faraday cage' is concerned, this effectiveness is not linked in any way to the presence or otherwise of earthing. That is to say the earthing of a DPR casing has no bearing on the effectiveness of the insulating effect of the casing. On the other hand, if any interference signals should enter the electronic DPR circuits that are housed inside the casing, along the cables, then how can earthing the casing prevent damage from this interference (especially from differential interference?). The answer is obvious: it cannot! Furthermore, it could be

Fig. 4.5 A printed DPR board with silver coated conductors (numbers 1 and 2), which are in contact with the earthed casing via a special spring.

argued that the earthing of a DPR cabinet only serves to exacerbate the situation and reduces the resilience to interference in protective relays. Thus in accordance with IEC Standard 60225-22-4 all the input and output circuits in a protective relay, with the exception of digital communication ports, are tested using a nanosecond pulse voltage with an amplitude of 4 kV. That is to say it is presupposed that these ports and circuits would not be able to withstand these tests. However, when a standard twisted pair cable is used or if these circuits are linked to an Ethernet network in place of the fibre optic cables, a high voltage would inevitably be applied to these circuits as in the situation illustrated in Fig. 4.4. How would the situation change if the DPR casing were to be carefully insulated from the earthing system? If they can be bypassed by stray capacitance (and in the example given below of a design variant they can actually be bypassed in this way) then on the basis of Fig. 4.4, high voltages would not be applied to these digital communication ports.

There is another problem with the existing earthing system in use today – HEMP, specifically its so-called 'fast' component – E1, which is characterized by a short but very powerful, electrical field pulse near the Earth's surface with a voltage of up to 50 kV/m and with a raising edge of approximately a few nanoseconds and a falling edge of approximately 1 μs [4.3]. This field has a complex structure and contains a vertical and horizontal component, which is responsible for the advent of considerable current pulses in long distance conductors, specifically in earthing systems, which act as large antennae absorbing electromagnetic energy over a wide area. In the event of a lightning strike or a spark gap failure in high voltage equipment fitted with functionally earthed elements (such as the earthed neutrals of Y-connected windings in high voltage transformers) the earthing system acts as an electrode with zero potential. In the majority of procedural documents, even ones as serious as in [4.4] no distinction is made between the threat to an earthing system from a lightning strike or the E1 HEMP component. For example, the following is taken literally from one of the documents [4.4]: 'Since the effect of EMP vectored interference is similar to that observed during lightning strikes, the lightning conductor system and the system of earthing electrodes are the main interface for the EMP protection system.'

However, there is indeed a significant difference between a high voltage lightning strike on an earthing system with zero potential or an insulation breakdown to ground potential in high-voltage equipment, and a powerful E1 electrical field pulse, part of which is directed parallel to the Earth's surface (that is to say parallel to the earthing system grid). In the advent of HEMP, the earthing system ceases to act as a zero potential surface and starts to act as a source of a powerful pulse voltage that is applied to electrical equipment, which is earthed in different parts of the earthing network and which are linked via a galvanic coupling (Fig. 4.4). Since this concerns both a very powerful and a very short pulse (that is to say with high-frequency characteristics), which creates a field strength in the air that can reach 50 kV/m it becomes clear that a large voltage differential can occur even in a small section of the standard earthing system, that considerably exceeds the value observed when voltages from a bolt of lightning pass through the earthing system. Therefore, the requirements for the electrical resilience of the insulation in all the input and output circuits in a DPR, that they be capable of withstanding pulse voltages in the nanosecond range with an amplitude of 4 kV, that are indicated in the IEC Standard 60255-22-4 are obviously insufficient to ensure the serviceability of a DPR. Apart from that it is no accident that we mentioned above that the a DPR casing was an element that is 'designed to act as a Faraday cage' as opposed to 'acting' as a Faraday cage. The reason for this is that metal casings for modern DPR are not effective as Faraday cages because of the presence of large recesses built into them to accommodate screens, keypads, and terminal modules (Fig. 4.6).

The parameters of the E1 HEMP component are such that all these recesses in the metal casing allow a high power electromagnetic wave into the DPR casing that has an equivalent frequency running into hundreds of gigahertz.

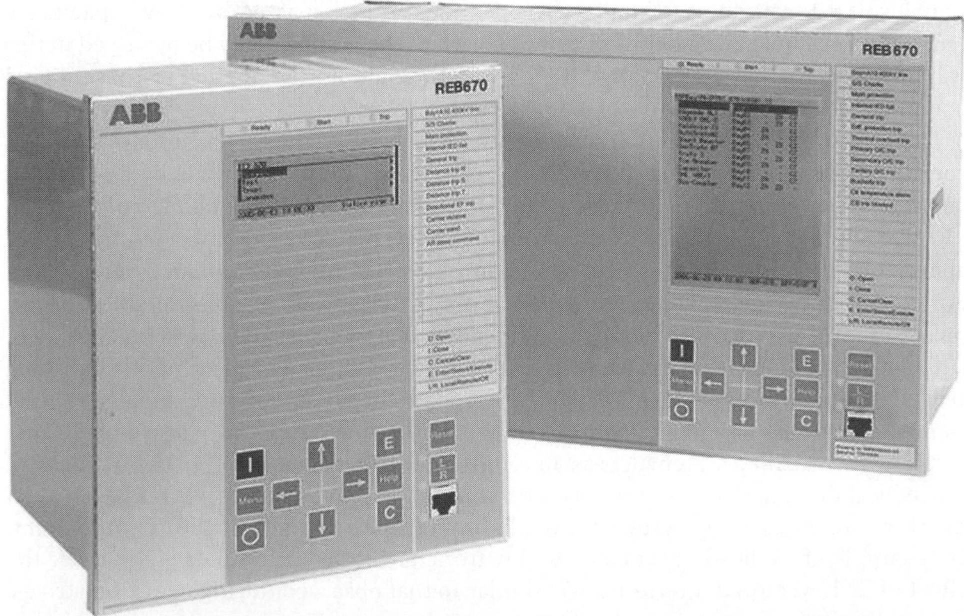

Fig. 4.6 Modern DPR terminals with a number of windows, recesses, and apertures for screens, buttons, indicator panels and other elements.

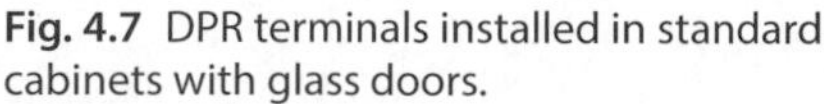

Fig. 4.7 DPR terminals installed in standard cabinets with glass doors.

Standard metal cabinets in which protective relay systems are accommodated today are also poorly suited to protecting DPR against high frequency electromagnetic fields since the underside is completely open (the same goes for the top) to enable entry for a number of cables, as well as glass doors through which the screens and indicator panels can be monitored conveniently without opening them, see Fig. 4.7. Therefore, one way or another an alternative solution has to be found to provide this protection. Thus it is clear that protective earthing is really necessary for DPR casings to protect personnel from electric shocks when coming into contact with this same casing, but functional protection is not necessary at all.

As far as protection from the threat of the E1 HEMP component is concerned, it appears that the known technical solutions for earthing systems that are applied in electrical power engineering are not only useless because of their high resistance with an equivalent frequency of dozens of gigahertz but are also a danger to sensitive electronic equipment. Thus the requirements for the earthing of DPR are contradictory to the requirements for ensuring their resilience to threats from HEMP.

According to [4.2] functional earthing cannot be examined separately to protective earthing, without violating health and safety standards. Let us doubt the truth of that statement and examine these two types of earthing separately and independently of one another. Using this approach provides an opportunity to arrange the earthing of DPR based on a new principle, which is founded on IEC recommendation 60364-5-584 [4.5] on enhancing the resilience of information technology equipment to interference by separating this equipment from sources of disturbance.

Since in this worked example the 'source of disturbance' is functional earthing, our proposal is to separate the DPR from this source, see Fig. 4.8.

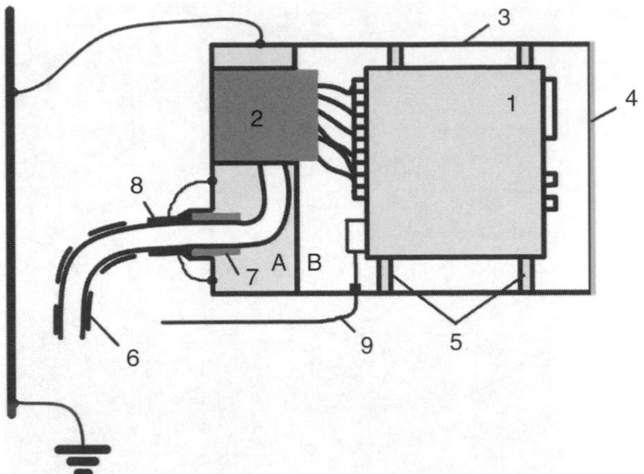

Fig. 4.8 The proposed principle for the layout of a DPR, which ensures enhanced resilience to all types of electromagnetic threats, including HEMP. A – 'contaminated' compartment; B –'clean' compartment, the DPR terminal in a carefully insulated plastic casing; 2 – HEMP filter; 3 – steel casing; 4 – door in the steel casing; 5 – insulators; 6 – control cable with twin screens; 7 – bushing insulator; 8 – a metal coupling to link the cable braiding with the steel casing and 9 – fibre optic cable.

In accordance with this proposal, the steel container 3, see Fig. 4.8, with the minimal number of apertures has been divided by an internal bulkhead into two zones: A – the 'contaminated zone' and B – the 'clean zone'. The DPR terminal in the plastic casing has been accommodated in the clean zone, which is free from electromagnetic emissions. The container has been fitted with a door 4, which provides access for personnel to the front panel of the DPR during prophylactic work. The container 3 is earthed and conforms to all the traditional norms and rules concerning earthing, which ensures the observation of the safety requirements. Given a sufficient distance between the DPR and the internal walls of the earthed metal container, of 5–7 cm for example, the stray capacitance from the electronic circuits in the DPR down to earth will be very insignificant and its influence can be dismissed. As far as the DPR casing itself is concerned, this should be carefully insulated (with plastic) and additional measures should be taken to prevent the efflux of dangerous voltage onto the surface of this casing. These measures could be: closing the screen with an additional transparent plastic panel; moving the control buttons onto the surface of the casing via the insulation spacers; powering the lights on the control panel located on the surface of the casing with light emitting diodes (LEDs) via fixed plastic clear rods; using an insulated optical port to connect an external computer to the DPR. Overall it is these simple approaches to ensuring security that are acceptable in the absence of earthing in hand-held electrical tools with so-called reinforced insulation and are not especially complicated to use in practice.

As far as the removal of a possible electrostatic charge, which could accumulate on the insulated DPR casing is concerned this problem can be solved by applying a thick, high-resistance, semiconductor coating onto the external surface of the plastic casing via a special high-voltage (50–100 kV) high-resistance (around 50 MΩ) resistor. An electrostatic charge would discharge to earth via a resistor such as this. The technology for applying this coating is already well advanced and is used widely in modern electronic apparatus. Compact high-resistance resistors operating on a voltage of between 50–100 kV are not hard to find and are produced by many companies, such as Caddock Electronics, Arcol, Ohmite, Welwyn Components and so on.

Another solution is based on the fact that the time staff spend working on an activated DPR is incommensurably short compared to the total DPR operating time. Therefore, the resulting idea of providing temporary grounding of DPR for when staff are working on them directly is an entirely reasonable one.

From a technical point of view, this idea can realized easily by so called 'position' (otherwise known as 'end' or 'limit') switches, installed on the relay cabinet door, to provide grounding of DPR enclosures and controllers as the door is opened. Modern position (or limit) switches are very reliable devices, and are well protected against mechanical failures as well as from adverse environmental impacts and are widely available on the market. They are extensively used in critical industrial systems, in road and air transport, military hardware and in systems designed to ensure the safety of personnel. The various different types of these switches can changeover currents of 10–16 A, at voltages of 400–690 V and can vary in terms of their complexity: from one NO (NC) contact to several groups of changeover contacts. In order to improve the reliability of the DPR enclosure grounding as the relay cabinet door is opened, two limit switches with contacts connected in parallel can be installed. To increase the electrical strength of contact spacing in the event of exposure to an electromagnetic pulse, switches rated to the maximum operating voltage (660–690 V) must be used. If there are two similar internal contacts available on the switch, they can be connected in series.

If permanent grounding is replaced with a connective grounding, all the metal enclosures of the DPR must be connected using flexible copper ties 2 to the common metal element 3 (see Fig. 4.2) located in the cabinet and insulated from it using small plastic cylindrical insulators. This common element should be connected to the cabinet grounding bus via the limit switch contacts that are closed as the door opens. In specific cases, if it is deemed reasonable, the use of additional grounding may be permitted ahead of any work in the relay cabinet: so-called temporary grounding. This grounding is achieved by manually linking by means of an insulated wire and a detachable contact (such as a powerful crocodile clip with insulated grips) the aforementioned element 3 with the cabinet grounding bus for the duration of any work in the open cabinet. Naturally the DPR enclosure must be insulated from the relay cabinet with a minimal capacitance and equipped with discharging high-ohmic resistors (as in the first instance). Ordinary DPR in standard enclosures can be used for this grounding method.

An additional solution is based on the so-called *paradoxical grounding* method. Today all grounding circuits are designed in such a way as to provide the *minimal impedance for a short impulse current.* Special equipotential surfaces are created in the control cabinets containing the electronic equipment for this purpose. Connection of these casings containing this equipment to these equipotential surfaces is carried out using special short, wide and flexible copper ties, see Fig. 4.2. That is to say all the necessary measures are taken so that the high power interferences from the grounding system can penetrate sensitive electronic circuits freely. However, in developing the idea set out here concerning the absence of the need for grounding of sensitive electronic circuits, it is possible to come to the paradoxical conclusion that grounding of the metal cases containing electronic equipment should be carried out using an opposite paradoxical principle. Such 'paradoxical grounding' should ensure the safety of personnel in the event that dangerous potential either from direct or alternating 50 Hz voltages should manifest itself on the equipment casing, but should block the power pulse interferences from penetrating the equipment from the grounding system. The realization of such *paradoxical grounding* is very simple and can be achieved by *increasing the impedance of the conductors connecting the electronic equipment with the common grounding bus for pulse currents.* For this purpose, a high-frequency choke with a high level of attenuation for short pulse interference can be inserted into the connection between the equipment enclosures and the grounding bus (as per our example in Fig. 4.2 – between the element 3 and the common ground bus in the relay cabinet). This choke does not have any influence on a direct current or an alternating 50 Hz current and provides reliable grounding for safety. Of course, the standard metal DPR enclosure must be insulated from the relay cabinet with minimal capacitance.

In our opinion, these proposed technical solutions enable the provision of a high degree of resistance to interference for DPR both in the natural operating conditions that actually exist today and in extreme conditions under the influence of HEMP or other destructive remote electromagnetic threat technologies [4.2]. Furthermore, the expenditure on the realization of this proposed technical solutions would not be overwhelming for electrical power systems. They could even be considerably less than the expenditure on the reconstruction of the old grounding system on many different electrical power sites, which would not permit DPR to function given the existing operating conditions.

4.3 HEMP Filters

The principle method of protection from the effects of HEMP on highly sensitive equipment is the careful electromagnetic screening of the apparatus itself together with the external cables attached to it and suppressing the pulse by using special filters.

4.3.1 Ferrite Filters

The simplest type of filter that does not require significant expenditure but which nevertheless is able to drastically weaken a short EMP (that is to say one which is analogous in terms of its properties to a high-frequency signal) in conductors that are attached to electronic apparatus is a ferrite filter made in the form of a ring (as a cylinder) that is placed onto the cable without the need for cutting, see Fig. 4.9.

The impedance of a coil, which is formed by one or several control cable windings, passing through the ferrite ring, is very weak for low frequency operating signals and for alternating current at industrial frequencies and very strong for high-frequency (pulse) signals within a specific frequency range, depending on the number of windings, as well as the material, and the geometric size of the ring. As a result, any pulse and high-frequency interference that entered this cable would have been significantly weakened. The weakening induced by these filters is 10–15 dB.

Many companies produce the majority of different types of filters, such as miniature examples intended for installation inside a piece of equipment on printed boards (Fig. 4.10) as well as installation directly onto wiring (cables). In order to make installation easier these filters are often produced in the form of two matching halves of a ring (semi-cylinders) located in plastic casings fitted with fastenings that ensure a fast and convenient installation of the filters onto the cables, see Fig. 4.11.

Fig. 4.9 Ferrite elements (FE) in filters.

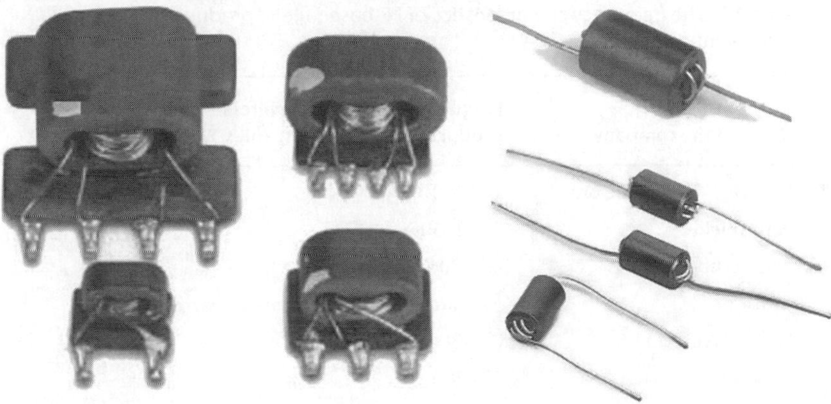

Fig. 4.10 Miniature filters based on ferrite elements (FE) and intended for installation on a printed board.

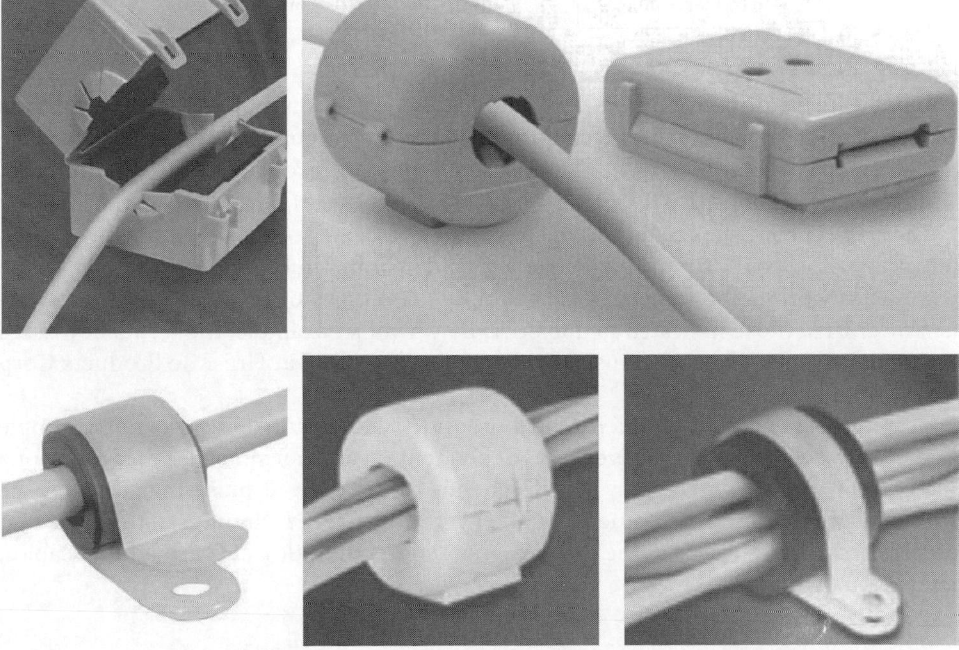

Fig. 4.11 The design of ferrite filters for quick and convenient installation on cables.

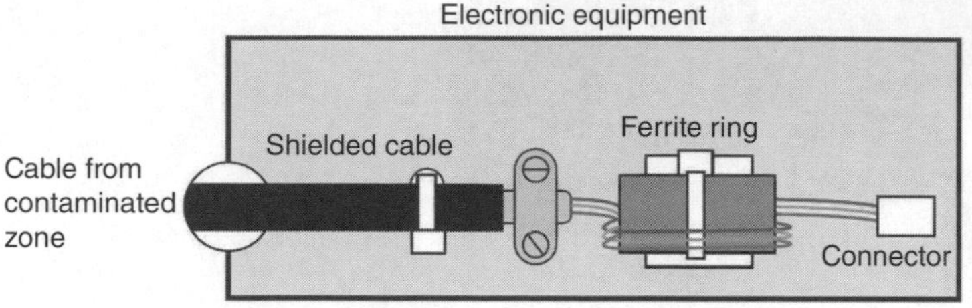

Fig 4.12 The installation of filters based on a ferrite ring onto a control cable entering a DPR.

Table 4.1 The frequency characteristics of FE-based filters produced by different companies.

Name of the company	Frequency range for the filters produced by these companies, MHz
Fire-Rite Products Corp.	1–1000
Ferrishield	30–2450
Ferroxcube	0.2–200
Murata	Miniature filters for printed boards
NEC/Tokin	0.1–300
Parker Chomerics	30–200
Laird	30–2000
TDK	10–500
Leader Tech, Inc.	1–2450
Wurth Elektronik	Miniature filters for printed boards

These filters can be used ubiquitously in relay protection: both in feed circuits, as well as in logic signal transfer circuits, as well as in the secondary current and voltage transformer circuits, see Fig. 4.12. FE-based filters are produced by many different companies, see Table 4.1.

The frequency ranges set out in Table 4.1 do not relate to any specific type of filter, but only serve to illustrate the frequency field within which one or another of the companies works. The frequency range for specific types of filters in reality is much narrower than those indicated in Table 4.1. By way of an example, the frequency range for different types of materials used in the production of FE at Fire-Rite Products Corp is set out in Fig. 4.13.

Despite their apparent simplicity and low cost ($1–10), ferrite filters are not as simple as they may seem. Their effectiveness is dependent on an enormous number of factors: the type of materials, the equivalent frequency of the current pulse that needs to be weakened, the geometric parameters of the FE, the number of windings in the cable that passes through it, the size of the direct current component flowing through the cables, temperature and so on.

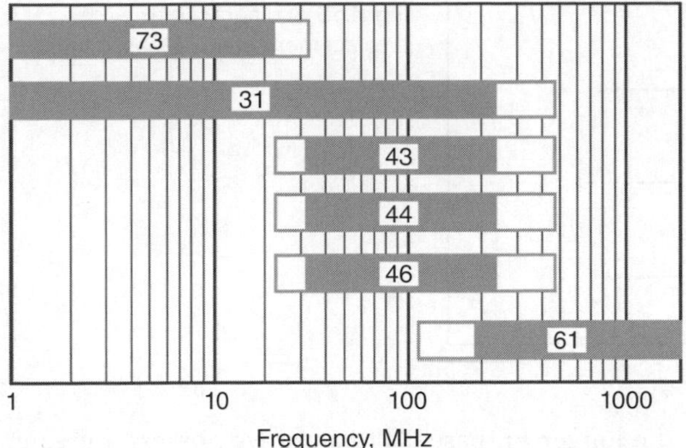

Fig. 4.13 The frequency ranges of different types of materials (defined by the numbers) used in the production of FE by Fire-Rite Products Corp.

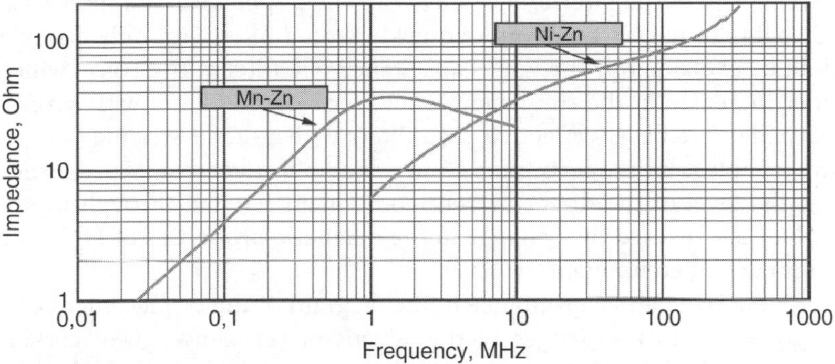

Fig. 4.14 The interrelationship of the impedance of an FE-based filter and the type of material used and the frequency.

The frequency characteristics of the filter are dependent on a number of parameters, first and foremost on the type of FE materials. For a frequency range of between 0.1–2 MHz, as a rule, manganese-zinc ferrites (Mn-Zn) are used with a magnetic permeability of μ = 600–20.000 while for a frequency range of 1 MHz – 2.45 GHz nickel zinc ferrites (Ni-Zn) are used with a magnetic permeability of μ = 15–2000. During the production process different ferrite blends are used.

Aside from the frequency characteristics the most important parameter for an FE-based filter lies in its impedance, which also defines the degree of suppression of interference the filter can offer.

The impedance of an FE-based filter is defined to a large extent by the type of material used as well as by the operating frequency, see Fig. 4.14.

Since FE-based filters possess inductivity, capacitance and active resistance, see Fig. 4.15 it follows that the filter's frequency characteristics and impedance are also dependent on the geometric dimensions of the FE, specifically its length, see Fig. 4.16.

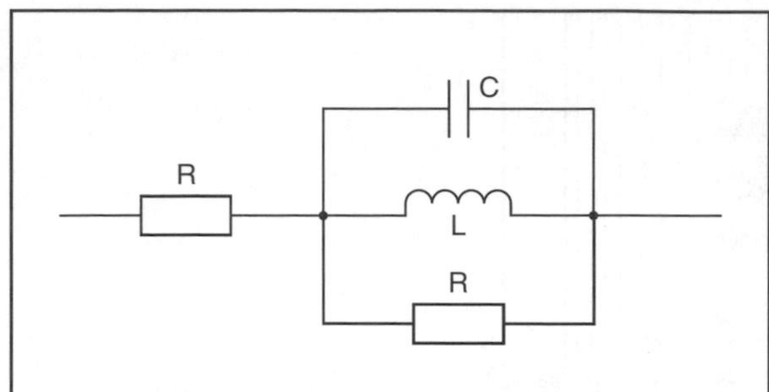

Fig. 4.15 A diagram of the displacement of an FE-based filter.

As can be seen from Fig. 4.16, the longer FE-based filters always possess a greater impedance all other things being equal, which explains the high inductive resistance of the filters with a longer FE.

The impedance of FE-based filters to a large extent also depends on the number of windings in the cable that passes through the FE, see Fig. 4.17. As is evident from Fig. 4.17, as the frequency of the interference increases the impedance of a filter with several windings begins to tail off much more quickly than a filter with only a single winding, this could be explained by the large capacitance of a filter with several windings. Given further increases in the frequency of interference the filters with several cable windings turned out to be less effective, than filters with a single winding.

Another relatively unpleasant property of an FE filter lies in the fact that its properties are dependent on the level of the direct current component flowing through it, see Fig. 4.18. This influence is caused by a change in the magnetic properties of FE in the presence of a direct current component.

The presence of inductivity and capacitance in the diagram of the displacement of a filter (Fig. 4.15) gives rise to the danger of the advent of resonance given certain frequencies, when instead of the interference signal weakening it increases, which is another unpleasant property of a filter such as this.

How should the correct filter be chosen to provide effective protection from HEMP given such a large number of factors influencing their parameters? This is not simple, especially if the lack of standards describing the procedure for selecting the parameters for these kind of filters and the fact that the different manufacturers are using different methods to obtain these measurements are taken into account. This makes the parameters of different filters produced by different companies practically incomparable with one another.

Based on this analysis it is possible to recommend the following basic principles for choosing the correct FE-based filter.

1) In order to effectively suppress pulse interference across as wide an HPEM frequency range as possible it is necessary to use three, as a minimum, filters placed in series in a single lead (cable) and which have been manufactured from different materials providing the maximum value for the impedance across all the filters that should lie in the low frequency (0.1 MHz), medium frequency (300–500 MHz) and high-frequency (2–2.45 GHz) ranges. The use of three filters placed in series in a single cable also solves the problem of resonance, since three filters each with different characteristics would have very different resonance frequencies.

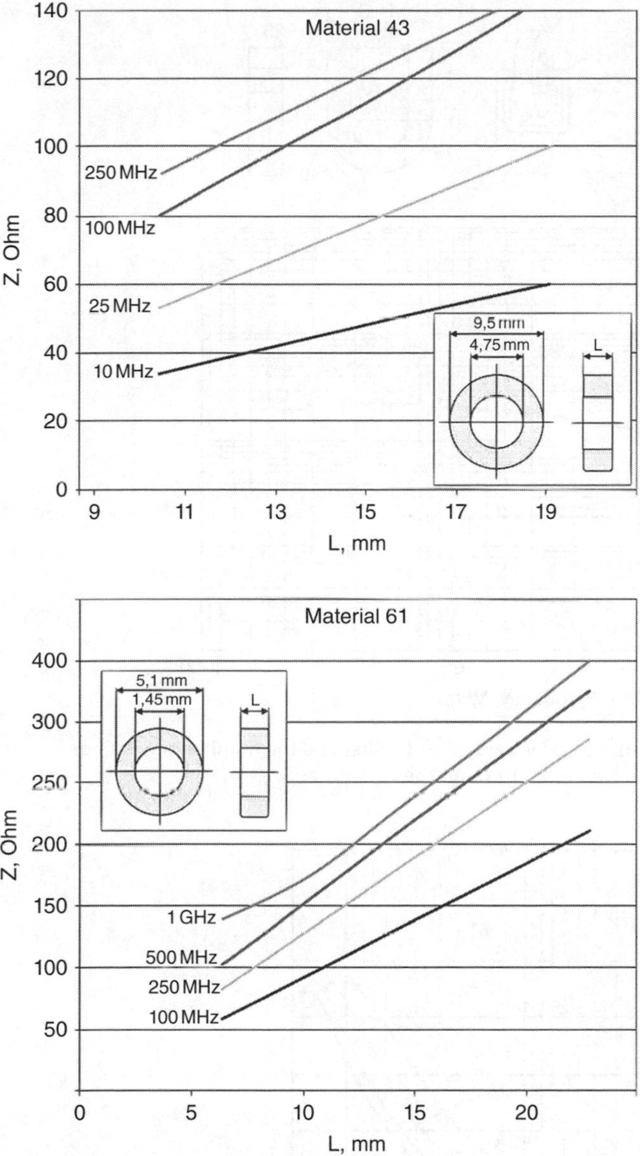

Fig. 4.16 The dependence of the impedance Z of a filter on the length L of ferrite elements produced from two types of materials (43 and 61) by Fire-Rite Products Corp.

2) The manufacturer's data could be used solely for the preliminary selection of the type of filter, after which a test of the efficiency of chosen filters at suppressing interference should be carried out across all the frequencies and currents of interest to the consumer.

A test such as this could be carried out at an installation containing an interference generator with realistic parameters (at the very least with a realistic frequency range)

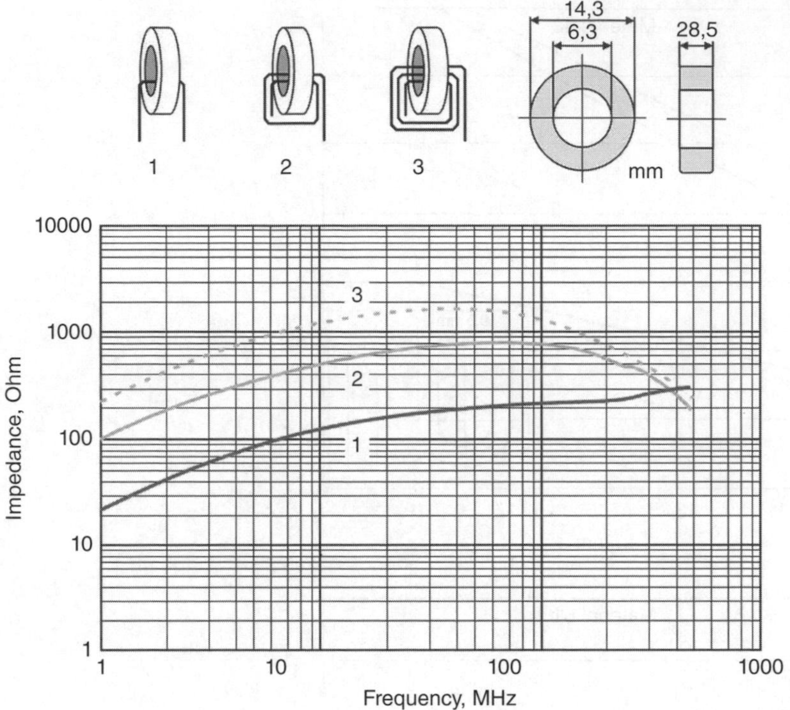

Fig. 4.17 The typical interrelationship of the impedance of a filter and the number of windings (denoted by the diagrams 1–3) to have passed through the FE.

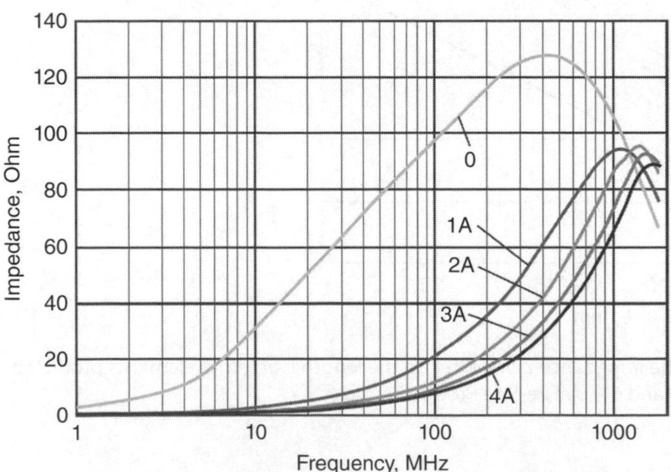

Fig. 4.18 The influence of the direct current component flowing through a filter on its characteristics.

and a receiver, which could be an oscillograph, a spectrum analyser or even an electronic voltmeter with an enhanced frequency range. The generator would be linked to the input terminal on the receiver using a cable with filters installed on it, see Fig. 4.19. It is very good practice to use a hand-held multifunctional device such as FieldFox

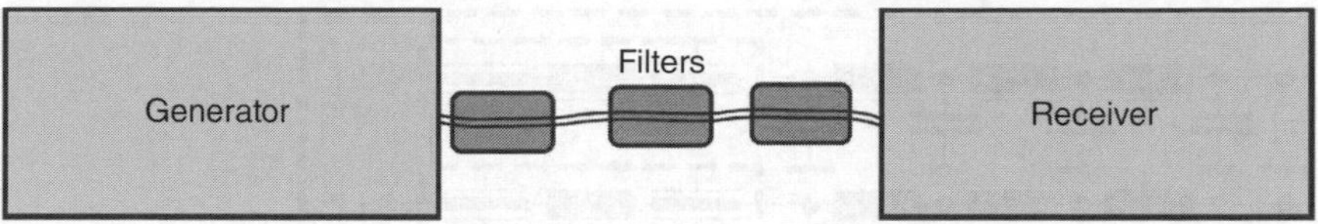

Fig. 4.19 An assembly designed to test the effectiveness of FE-based filters.

Microwave Network Analyzer NN9918A type (Keysight Techn.) that contains a generator, a receiver and spectrum analyser in a single small unit and can measure the attenuation of a filter directly.

4.3.2 LC Section-based Filters

Many companies also produce special LC section-based filters (see Fig. 4.20) through which the electronic equipment being protected, and the external devices and systems are linked together.

The contemporary market for filters such as these is represented by dozens of different types, manufactured by a number of companies: ETS-Lingdren, MPE, Meteolabor-EMP, European EMC Products Ltd, Captor Corp, LCR Electronics, API Technologies, Astrodyne TDI Corp, Fi-Coil, EMI Solutions Pvt. Ltd, RFI Corp and so on. It would appear that the problem is as follows: would you like to protect your infrastructure from HEMP? Then install filters and you can sleep soundly! The question, however, is this: Can you really sleep soundly after fitting these filters?

Any attempt to select a filter that is capable of suppressing HEMP effectively suddenly comes up against a problem: all the companies listed previously advertise their own filters as highly effective protection against HEMP and furthermore they claim their filters are compliant with the military standard MIL-STD-188-125 [4.6]. However, this being the case they display parameters under test pulses that would be sustained by filters that differ considerably from those listed in the standard. Thus this standard, for example, supposes that the filters have been tested using current pulses with a specific amplitude of 20/500 ns for a load factor of 60 Ω while these filters are tested by manufacturers using current pulses of 8/20 ms for a load factor of 1 Ω. So why is this the case? This is because the 8/20 ms pulse is the standard pulse produced by all types of testing apparatus designed to test resilience to a lightning strike, while specialized and very expensive apparatus is required to test using current pulses of 20/500 ns for a loading of 60 Ω for which the manufacturers do not possess the filters. This problem is discussed in [4.7].

Another peculiarity is that the MIL-STD-188-125 supposes that the filter has been tested using a current pulse in two modes: in the common mode – where the current flows through all connected together inputs and down to earth, and the wire-to-ground mode – where it flows through each input in turn and then down to the ground.

However, during a high altitude nuclear explosion, a high-voltage pulse can be applied not only through the input points on a piece of apparatus and the Earth (in other standards this mode is described as the 'common mode') but between different input points within a piece of apparatus that has been insulated from the earth ('differential mode').

The standard MIL-STD-188-125 does not envisage these tests as they are examined under a different standard. In connection with this several companies that advertise their filters as HEMP filters do not fit them with pulse voltage limiting elements at all

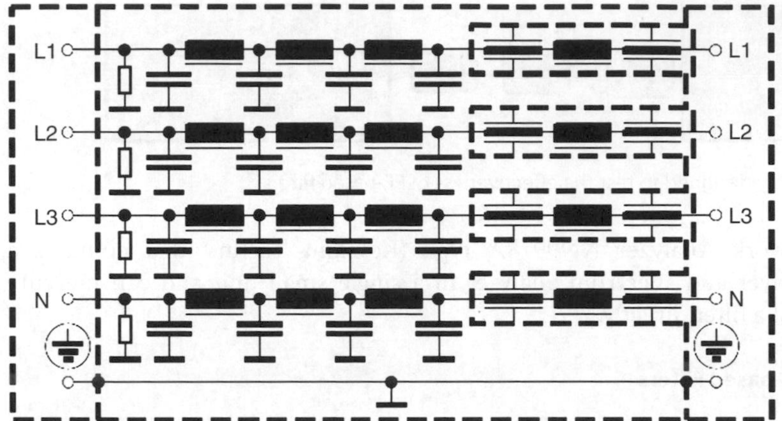

Fig. 4.20 A typical design of a high-power HEMP filter, consisting of a set of LC section filters.

and everything is linked to that same MIL-STD-188-125. Furthermore, they confirm that since their filters have undergone testing using pulse currents with an amplitude of several thousand amps and are recognized as complaint with MIL-STD-188-125 this means that they provide comprehensive protection from HEMP. In reality, under this standard, pulse voltage limiters are presented as completely independent elements, which do not bear any relation to the filters, see Fig. 4.21.

Given this approach, the filters do not have to provide any protection from overvoltages at the input points. This being the case is it possible to confirm that these same filters provide comprehensive protection from HEMP? It is obvious that in their understanding of this problem many manufacturers fit their filters with elements installed at the input points that are designed to protect them from pulse overvoltages. In their opinion, these filters can rightly be called HEMP protection filters.

However, a careful examination of the protective elements used in these filters raises serious doubts concerning their effectiveness. The most widely used and the cheapest pulse voltage limiters used in filters are gas discharge tubes and zinc-oxide varistors, see Fig. 4.22. It is well known that they are relatively 'slow' elements, which cope well with suppressing standard pulses of 8/20 ms but are not able to function under the influence of a short high-voltage HEMP pulse with parameters of 2/25 ns [4.8] (or 5/50 ns according to other data [4.9]). The consumer, on reading that their chosen filter is designed to protect from HEMP and that it has undergone testing with a pulse current of 8/20 ms and is fully compliant with MIL-STD-188-125 would be unlikely to look up this standard and check if it was actually this same pulse.

However, it is too early to send complaints to the manufacturers. In [4.10] it is stated that the relatively fast varistors installed in the filters and the slower GDT, it turns out, are not designed to protect from HEMP at all but are protective elements designed to protect from lightning and switching overvoltages. That's the way it is: the voltage limiters used in filters designed to protect from HEMP are themselves designed to protect from lightning strikes, but not from HEMP! Nevertheless, some of the filters produced by MPE and fitted with protective varistors are referred to in the marketing brochures

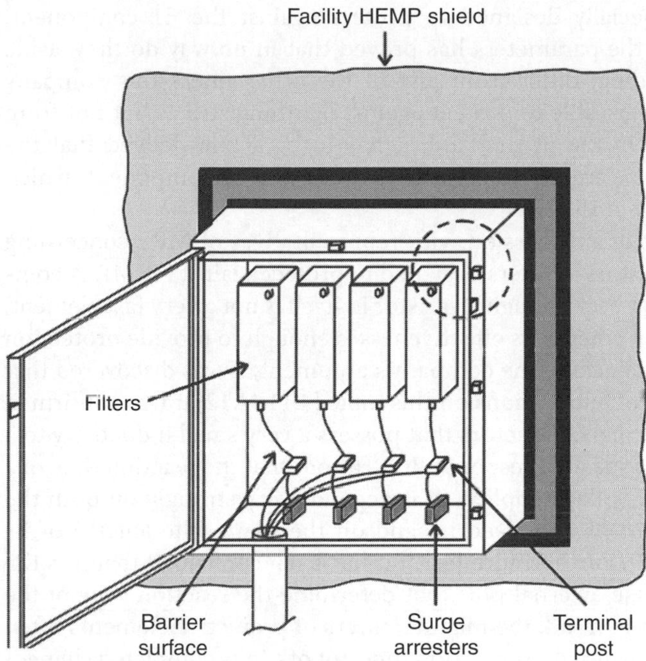

Fig. 4.21 An inlet box assembly for attaching an external cable to an asset protected from HEMP (according to MIL-STD-188–125).

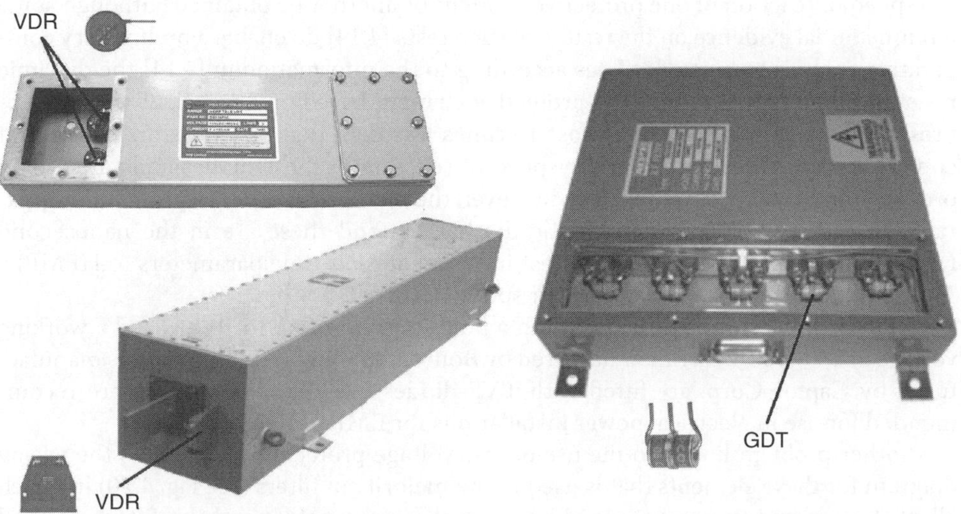

Fig. 4.22 Filters produced by MPE with protective elements at the input points, for which zinc-oxide varistors (or voltage depending resistors – VDR) and gas discharge tubes (GDT) are used.

as filters that have been specially designed to protect against the E1 component. However, careful analysis of the parameters has proved that in no way do they, aside from the name in the branding, differ from any of the other filters this company produces. That is to say they are able to protect against lightning strikes but not from the E1 HEMP component. On the other hand, it has to be acknowledged that the parameters of a lightning strike are no different to those of the E1 component, which does not correspond to reality at all.

A new argument came out of a discussion with representatives of MPE concerning the justification for using varistors in filters designed to protect against HEMP. A company representative stated that even though a varistor in itself is not a very fast element, in conjunction with LC filter elements its effectiveness is enough to provide protection from the E1 component. On checking the company's argument, it was discovered that this was incorrect. In a series of publications on this topic [4.11–4.12] it was confirmed that attaching even short external conductors that possess a very small inductivity to a protective element reduces its speed of response. It turns out that the reaction time of a protective element to a pulse voltage applied to it depends very strongly on both the design of the casing this element is housed in, and on the shape (the length) of its pin configurations [4.11–4.12]. Furthermore [4.12] it has been confirmed that it is the very design and length of these external pins that determine the reaction time of the protective element. With this in mind, the manufacturers of protective elements often indicate in their promotional material the reaction time not of a fully completed element with pins inside a casing, but just that of the material used to manufacture this same protective element [41.2]. In addition to this the manufacturers are working on an upgrade to the design of the outputs of their protective elements and are achieving some considerable success [4.13].

From the information set out here it is clear that only by conducting your own independent tests on manufactured items available on the market can any objective data on the speed of reaction of one protective element or another be obtained, although some circumstantial evidence on the results of such tests [4.14] do enable a preliminary comparative analysis to be made. Thus according to the information in [4.14] the dynamic resistance and reaction time of a protective element based on a transient voltage suppressor diode (TVS-diode) is almost 10 times less than that of a varistor. We did not conduct our own research into the speed of reaction of these diodes and varistors in order to prove or disprove this data, however, the fact that in order to protect the apparatus from high-voltage electro-static discharges (and these are in the nanosecond range, that is to say they are the closest in terms of their time parameters to HEMP) a TVS-diode is used instead of a varistor speaks for itself.

The most powerful TVS-diodes (with a pulse current of up to 10 kA, and a working voltage of 200–500 V) are manufactured by Bourns, Inc. and Littelfuse. Filters manufactured by Captor Corp. are fitted with TVS-diodes (see Fig. 4.23), which are recommended for use in electrical power installations for EMP protection.

Another problem linked to the use of overvoltage protective elements is the wiring diagram for these elements that is used in the majority of filters (see Fig. 4.20) in which all of these elements are energized between the input points and the filter's earthed casing. If the elements are wired in this way it would mean that two elements connected in series are energized between two input points on the filter, which causes a double discharge voltage that could pose a danger to the electronic circuit being protected.

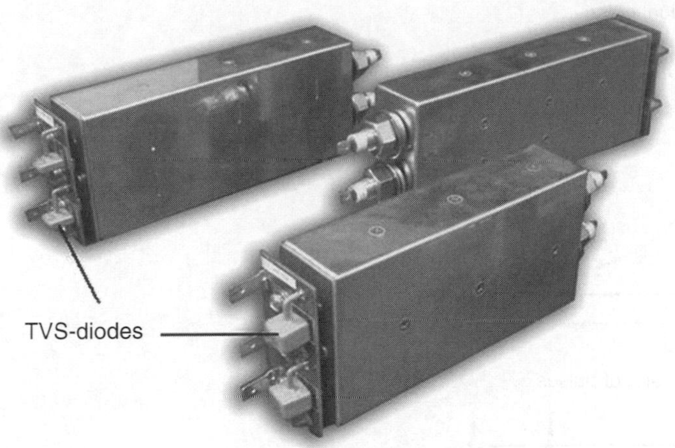

Fig. 4.23 HEMP filters produced by Captor Corp, and fitted with high-power high-speed TVS-diodes.

The technical requirements for the resilience of an apparatus to these currents and the methods of testing are described in the IEC 61000-4-4 [4.9] and IEC 61000-4-25 [4.15]. The test pulse voltage used was the so-called Electrical Fast Transient (EFT) – a fast pulse, the parameters and methods of testing for which are described in IEC 61000-4-4, see Fig. 4.24. The method by which the parameters of test pulse voltages are selected on the basis of these standards, to use a specific example, are as follows – a DPR, as set out in [4.16] for which the amplitude of the pulse voltage EFT is 8 kV.

In our opinion, the filters with pulse voltage protective elements that are designed to protect against HEMP should undergo these tests in addition to the tests envisaged in the standard MIL-STD-188-125 and, furthermore, with test voltages applied both between the input points and the earth and between two individual input points.

Another problem is linked to the amplitude-frequency response of filters. A typical response from a high quality filter designed to protect from HEMP is set out in Fig. 4.25. How are the responses of actual filters linked to this typical response?

One of the companies (such as Meteolabor for example) do not set out any data in the technical description for some of their filters concerning the frequency range and the attenuation that the filters are able to impart. Other companies (such as MPE) set out the typical characteristics in their documentation that the filters should possess if they are compliant with the MIL-STD-188-125 standard (a frequency range of between 14 kHz – 40 GHz and an attenuation across the whole range of 100 dB) and also set out the actual characteristics of their manufactured filters with the parameters of: 10 kHz – 1 GHz with an attenuation of 80 dB – for a filter of average quality and the same frequency range but with an attenuation of 100 dB for a high quality filter. What has happened though to a frequency range of between 1 and 40 GHz? Employees at Astrodyne (LCR Electronics, Inc.) turned out to be even more ingenious. In the technical documentation concerning their filters they wrote that they provide an attenuation of 100 dB across a frequency range of 14 kHz – 10 GHz in accordance with the MIL-STD-220 [4.17]. If their filters, however, were well insulated and shielded (read: installed inside a 'Faraday cage') then their frequency range could be expanded to the required

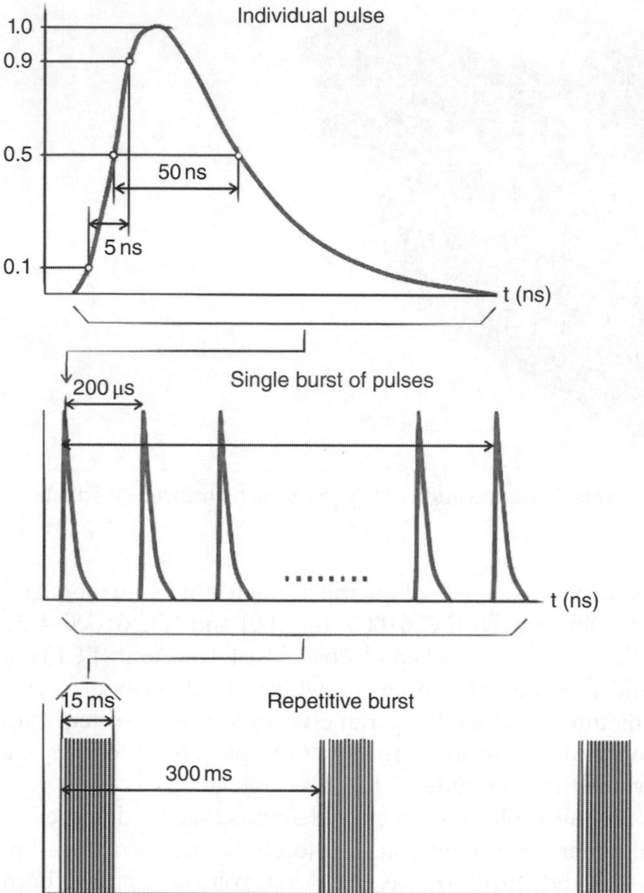

Fig. 4.24 The fast Electrical Fast Transient (EFT) pulse (IEC 61000-4-4).

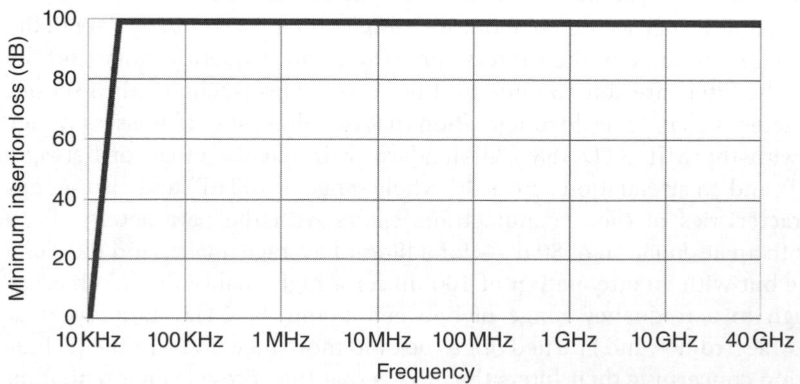

Fig. 4.25 A typical amplitude-frequency response characteristic for high-quality HEMP filters.

Table 4.2 The frequency characteristics of filters produced by leading manufacturers.

Manufacturer of the HEMP filter	Frequency interval		Attenuation, dB	Notes
	Min	Max		
LCR Electronics	14 kHz	1 GHz	100	–
MPE	10 kHz	1 GHz	80	For a standard filter
MPE	14 kHz	18 GHz	100	For an enhanced filter
Fi-Coil	14 kHz	1 GHz	100	–
Captor Corp.	14 kHz	10 GHz	100	–
ETS-Lindgren	14 kHz	40 GHz	100	–
MeteoLabor	200 kHz	1 GHz	80	For a PLP type filter
Meteolabor	–	–	–	For a USP type filter

value of 40 GHz. That is just the way it is: In order that a filter protect against HEMP 'correctly' it needs first to be protected from this same HEMP! As the saying goes, enough said.

The entire funny side of the situation is set out in Table 4.2, in which all the data concerning the frequency characteristics of the filters produced by leading manufacturers are set out for comparison: all the filters are designed for the same purpose and they are all compliant with the MIL-STD-188-125 standard and what is more all their parameters differ considerably. How then can this be?

Another idiosyncrasy of this wiring diagram in which all the internal filter elements are wired separately between each of the input points and the earth is not only the double discharge voltage on the protective elements from pulse voltages, as described previously, but the double capacitance and double inductivity activated between each of the input points in comparison with the inductivity and capacitance between the input point and the earth. From this it follows that the frequency characteristics of the pulse filters designed for pulses applied between the input points would not be the same as those for a pulse applied between the input points and the earth. How acceptable would these characteristics be then in terms of protection from HEMP?

What sort of consumer would actually 'drill down' into this? Why would they not believe the manufacturer's confirmation concerning the effectiveness of their product? Even if they did not believe it they are for the most part not in a position to check the actual effectiveness of the 'operation' of their filter independently. The effectiveness of a filter in protecting the equipment it is responsible for would only come to light if it were unable to provide protection from HEMP in an emergency situation.

Today, the situation is such that each manufacturer can decide for themselves if they would like to incorporate pulse voltage limiters into their filters or not; and whether or not to use elements that may be cheap but in terms of their parameters are not suitable; and whether or not to test their filters using a standard 'lightning pulse' or a pulse displaying the parameters specified in the military standard, whether to mention or not to mention the frequency range, or to use the frequency range specified in the MIL-STD-188-125 or the MIL-STD-220 standards [4.17], or to indicate their own frequency range and the make reference to the known military standard. Who is going to check?

The situation that has come about is a consequence of the lack of a specific standard that specifies the requirements for the design and the parameters of HEMP filters, the methods for their testing, and the criteria for their performance. In our opinion this is an unacceptable state of affairs, considering the weight of the problem, and it needs to be resolved immediately.

4.4 Non-linear Overvoltage Limiters

One effective method of combating overvoltages at the input points of electronic apparatus and its power supply terminals is to use filters with non-linear characteristics: gas dischargers, varistors, special semiconductor elements based on the avalanche effect and so on, that are connected in parallel with the asset being protected (e.g. in parallel with the input point of a DPR) and with each of its terminals and the Earth. Today, resistors with non-linear characteristics are considered the best performing elements, and are manufactured using pressed powder zinc-oxide ZnO (or more rarely from silicon carbide, barium titanate and other materials) – varistors that have been sold most widely. Today, varistors are manufactured in huge quantities: without casings, in different types of casings, and are often supplied with all kinds of auxiliary elements (fuses, signal flags etc.). Varistors should be selected carefully. Unfortunately, it is often the case that varistors are selected incorrectly, even in apparatus produced by leading global manufacturers, and in actual fact they do not have any protective function at all.

Since the VAC for a zinc-oxide varistor is far from ideal (see Fig. 4.26), it is not an easy task to select the correct one. On the one hand, the varistor should not allow a current of more than 1 mA to pass through it (the standard value for contemporary varistors) at the maximum operating voltage (otherwise it would just overheat and burn out); on the other hand, the residual voltage (clamping voltage) in this varistor should be an appreciably lower voltage than that which the equipment being protected can withstand

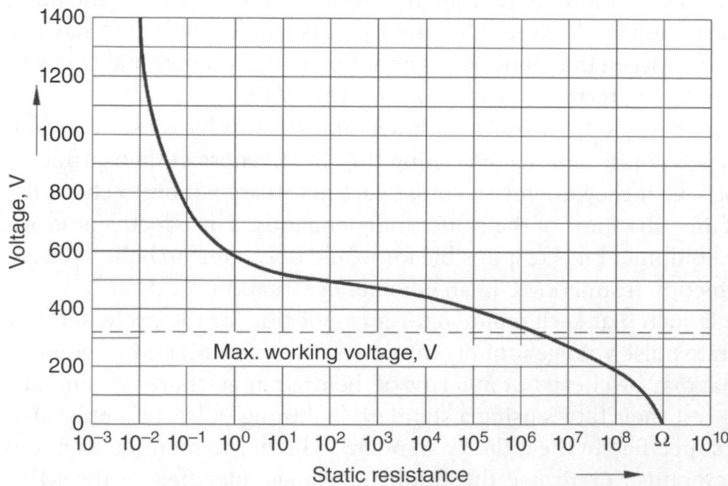

Fig. 4.26 The maximum Volt-Ampere Characteristic (VAC) for a zinc-oxide varistor.

(otherwise the varistor would not be protecting the electronic components, but the electronic components would be 'protecting' the varistor).

Owing to the inadequacy of the VAC for varistors as far as performing to these specifications is concerned, the maximum withstand voltage for electronic components designed to operate in a 200 V circuit, should be no less than 1000 V. However, firstly the electronic components that can withstand this voltage are much more expensive than lower voltage components, and secondly the other characteristics they possess are worse. For example, the 1000–1200 V transistors have a much reduced amplification coefficient and a significantly larger drop voltage than 400–500 V transistors. Thus it is often the case that transistors with a maximum withstand voltage of 500 V, that operate in a 220–250 V circuit are found in the power supply for digital fault recorders, DPR and in other electronic apparatus produced by leading global manufacturers. It is just not possible to ensure the protection of electronic components using varistors given this interrelationship of operating and maximum withstand voltage.

Contemporary metal oxide varistors are high-powered, see Fig. 4.27, and are produced in plastic casings, specially designed for installation onto a standard DIN-rail in mounting cabinets, see Fig. 4.28.

Unfortunately, these wonderful elements do not demonstrate a sufficiently fast reaction time (for protection against HEMP). Their characteristics deteriorate (are degraded) under the repeated impact of high-power pulse loadings.

Furthermore, the leakage current through the varistor increases in a normal voltage, and the varistor begins to overheat in which case its resistance is reduced but the current continues to grow, right up until it burns the varistor out. This can result in a short circuit in the supply network along with all the consequences that arise from it. Therefore, recently varistors have started to come onto the market with built in fuses, connected in series to the varistor. If the current passing through the varistor should increase sharply the fuse will burn out and activate a warning LED, see Fig. 4.29. Sometimes, in place of a standard fuse, a bimetallic contact located on top of the varistor itself is used, which is

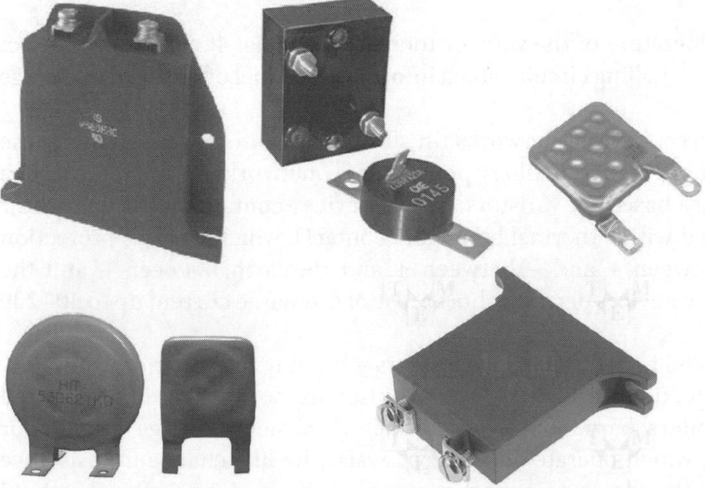

Fig. 4.27 Different types of high-power varistors with a nominal voltage of 130–1100 V and a discharge current of 3–100 kA.

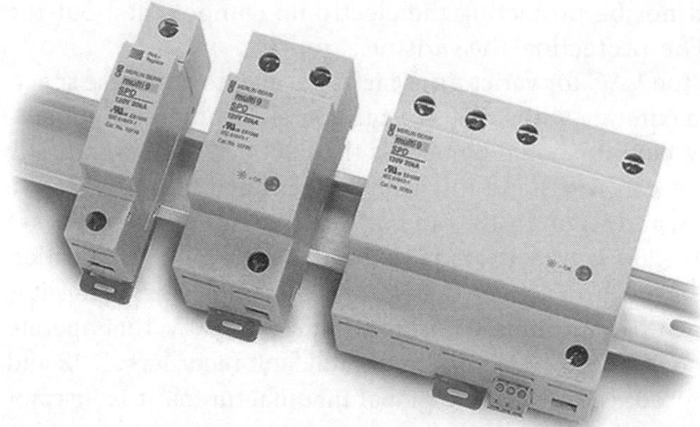

Fig. 4.28 High-powered protective devices based on metal oxide varistors produced by Square D (Schneider Electric), designed for installation on a standard DIN-rail.

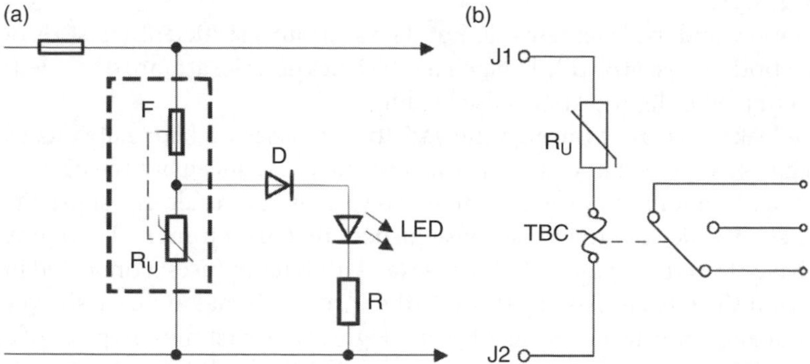

Fig. 4.29 A wiring diagram for varistors with a fuse F (a) and with a thermal bimetallic contact TBC (b).

activated when the temperature of the varistor increases, see Fig. 4.29(b). This contact can be connected to the signalling circuit, which informs personnel of the need to change the varistor.

In order to protect direct current networks (in substations and power stations these are multifaceted and long-distance auxiliary power supply networks), special protective devices are used that are based on varistors. These devices contain, as a rule, two or three varistors (equipped with a thermal bimetallic contact), which ensures protection from overvoltages of between '+' and '–', between '+' and 'the earth', between '–' and 'the earth', see Fig. 4.30. These devices permit a short pulse of discharge current up to 40–200 kA or more.

The shortcomings noted here concerning varistors – notably that their reaction time is deficient in their protection from short HEMP pulses are not present in high-speed silicon overvoltage limiters based on Zener diodes (Transient Voltage Suppressor Diodes – TVS-diodes), which operate on a sharp, avalanche like change in resistance from a relatively high value down to almost zero as a voltage of a specific threshold value is applied to them, see Fig. 4.31.

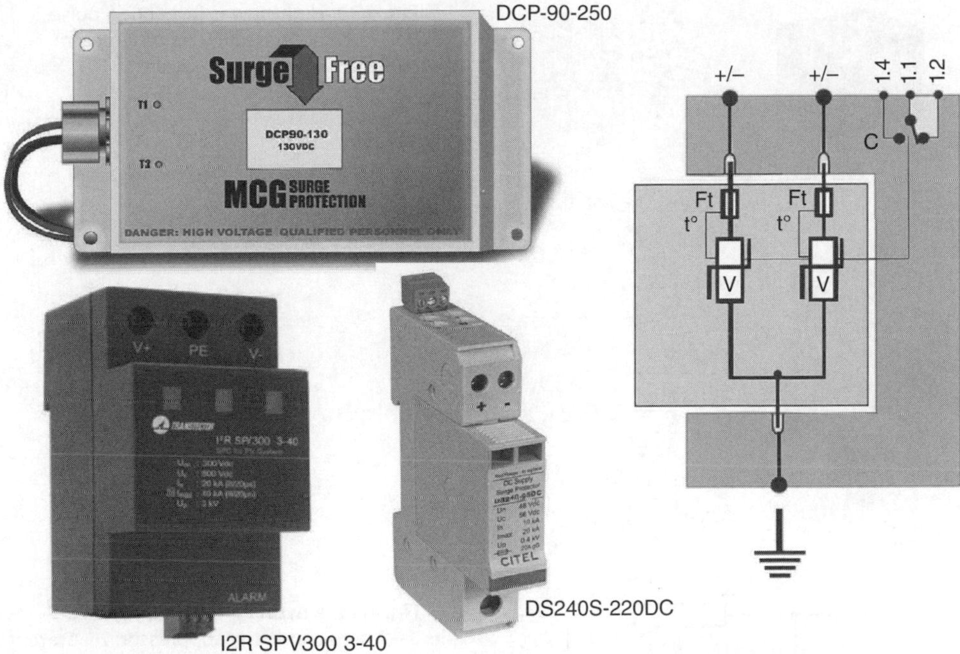

Fig. 4.30 Protective devices with varistors for direct current networks with a nominal voltage of 220 V.

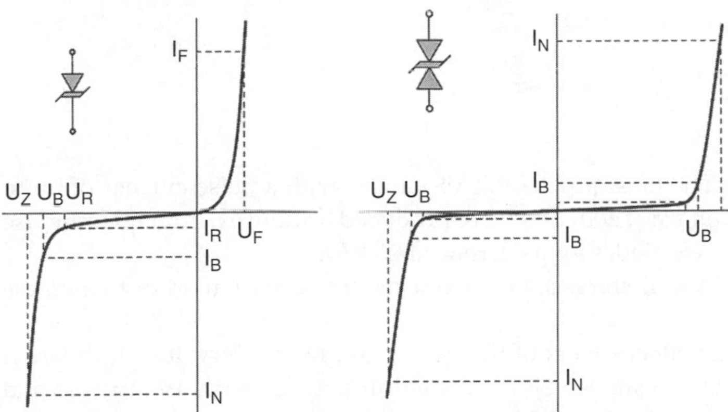

Fig. 4.31 The VAC of unidirectional (for direct current) and bidirectional (for alternating current) TVS-diode.

Apart from that, in contrast to varistors the characteristics of these TVS-diode based overvoltage limiters are not eroded as a result of the repeated impact of high voltages and mode switching. Up until recently, these protective elements demonstrated insufficient pulse power dissipation and were therefore used only in low power circuits in electronic equipment. However, high-power suppressors have recently begun to come onto the market, see Fig. 4.32. The TVS-diodes produced by UN Semiconductors are capable of passing a pulse current with an amplitude of up to 3 kA and operate at

Fig. 4.32 High-power, fast acting pulse voltage amplitude limiters based on avalanche diodes (TVS-diodes).

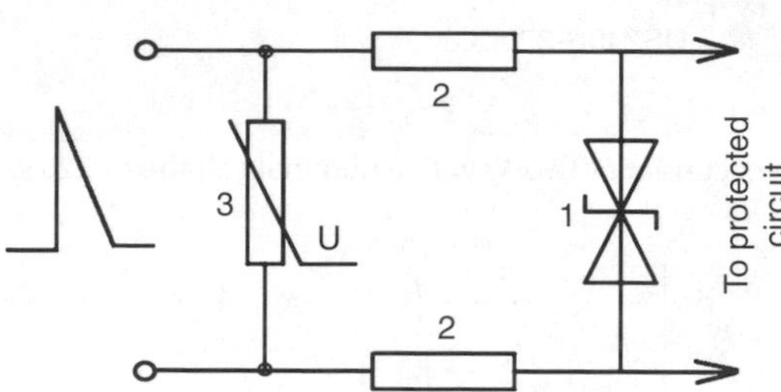

Fig. 4.33 A hybrid protective device: 1 – A semiconductor suppressor; 2 – Current limiting resistors; 3 – A high power varistor.

voltages of up to 440 V. The most powerful TVS-diodes with a pulse current of up to 10 kA and an operating voltage of 200–500 V are produced by Bourns, Inc. and Littlefuse (the company develops TVS-diodes for a current of 20 kA).

The TVS-diodes, just like varistors, can be connected in parallel in order to increase the discharge current.

In order to enhance the effectiveness of the protection from overvoltages, different types of overvoltage limiters can also be connected in parallel, such as varistors and semiconductor suppressors, see Fig. 4.33. A hybrid device of this nature possesses excellent characteristics.

The first to be activated in this hybrid device is always the fast acting suppressor 1, which even reacts to a pulse with a very steep leading edge and absorbs part of its energy. The discharge current is limited by the resistors 2, which prevents the destruction of the suppressor. A voltage drop in resistors 2 increases the voltage on the varistor 3, which leads to a sharp decrease in its resistance and in the shunting of the resistors. The remaining (greater part) of the energy is absorbed by a powerful varistor.

Once the high-power varistors have been installed the direct and alternating current systems busbars and the less powerful TVS-diodes have been installed immediately adjacent to the input points of the electronic apparatus being protected, this becomes a highly effective protection system.

4.5 Shielding of the Control Cables

The principle means of protection for the control cables from induced voltages lies in their shielding, together with a careful choice about where the cables are laid taking into account the maximum permitted distance from lightning conductors and from high-power cables, by using special cable trays. There are several different types of these cable trays: plastic with aluminium inserts, plastic with a metal spray coating and aluminium.

In general, the effectiveness of the metal shield (that is to say the degree of attenuation of the electromagnetic field) is provided by two of its properties: the absorption of energy as the electromagnetic wave passes through the conductive material and the reflection of the wave on the boundary of the division of the two materials. Both these materials depend both on the frequency of the electromagnetic wave and on the kind of material the shield is made from. The best absorption of electromagnetic energy is provided by ferromagnetic materials (such as iron, permendur, permalloy) and the best reflection of an electromagnetic wave is provided by diamagnetic materials (such as copper or aluminium). The effectiveness of the shielding properties of ferromagnetic materials is reduced as the field level increases due to its saturation, and the effectiveness of diamagnetic shields is reduced as the frequency increases due to an increase in the resistance. For a number of reasons that are technical and economic in nature shields with copper grid (braiding) and with different aluminium profiles have been used most widely.

Since the depth of the penetration of an electromagnetic wave into metal is inversely proportional to the frequency of this wave, then it is obvious that the thicker the metal shielding is the broader the frequency range over which it can attenuate the electromagnetic field. For example, if a copper shield with a thickness of approximately 0.6 mm is sufficient to provide effective shielding at a frequency of 500 kHz then for an industrial frequency of 50 Hz a copper shield would be required with a thickness for the walls of approximately 6 cm (a thickness of 5 mm is sufficient for a ferromagnetic screen).

From this it is obvious that plastic cable trays with a metal spray coating, which are widely used to shield control cables, demonstrate the minimal attenuation shielding effect. This design only begins to work effectively at frequencies of 600 MHz or more. At frequencies of less than 200 MHz it does not work at all. The interference entering the control cables in substations, for the most part, has a lower frequency than the 200 MHz indicated here so plastic cable trays with a metal spray coating are no use at all. Furthermore, aluminium cable trays and a copper braiding on the cables are still capable of attenuating induced voltages by a factor of 10 and as such they have been used widely. The greatest degree of attenuation of interferences with a broad frequency range comes from shielding control cables in steel water pipes.

In order for the shielding to function properly it is necessary to ensure that the charge that is induced towards it is discharged to earth. In an ideal situation the potential along the entire length of the shielding should be equal to the earth voltage, therefore, now and again in especially sensitive high-frequency circuits it is necessary to apply multiple earthing of the shielding cable every 2 λ (λ- is the wavelength for the electromagnetic wave).

When laying shielded cables in a substation an additional solution can be used in which a copper bus can be laid parallel to the route of the cable to equalize the

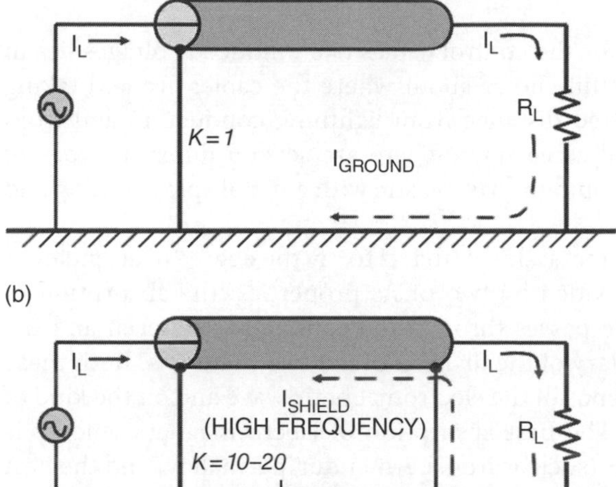

Fig. 4.34 The operation of a shield, earthed from one side and then from both sides.

potentials, which is earthed from both sides. However, for economic reasons a simple earthing of the shield is used from either one or two sides, see Fig. 4.34.

Often, one has to listen to the opinion of relay protection specialists talking about the expediency of earthing control cable shields from only one side. It is evident that this opinion has arisen out of two factors that are known to relay protection specialists: the earthing of current circuits from one side only and the earthing of high-voltage cables from one side only. These two factors sometimes mean that the control cables are earthed without taking into consideration the fact that in the examples set out here, earthing is a means of ensuring electrical safety and not protection from interference.

In reality the earthing of a control cable shield from the one side is only effective against capacitive interference, see Fig. 4.35. (so-called electrostatic protection) and is a completely ineffective measure (the interference reduction factor being k = 1) against inductive interference, since this shield does not provide loops for the interference current.

When the shield is earthed from two sides there is an additional circuit (shield), which possesses a considerably reduced impedance, for a higher-frequency signal than the earth. As a result, the operating signal is divided into two and one part (the low-frequency element) returns via the earth as before, while the second (the high-frequency element) via the cable shield. Thus for a high-frequency component the current inside the shield is equal to that in the central core, that is directed in the opposite direction and is balanced by the electromagnetic coupling between the shield and the central core. This ensures protection from the high-frequency emissions from the central core to the external space (that is to say to neighbouring cables) with an interference reduction factor of k = 3–20. This system operates just as effectively given an external electromagnetic impact on the shield, during which a high-frequency signal

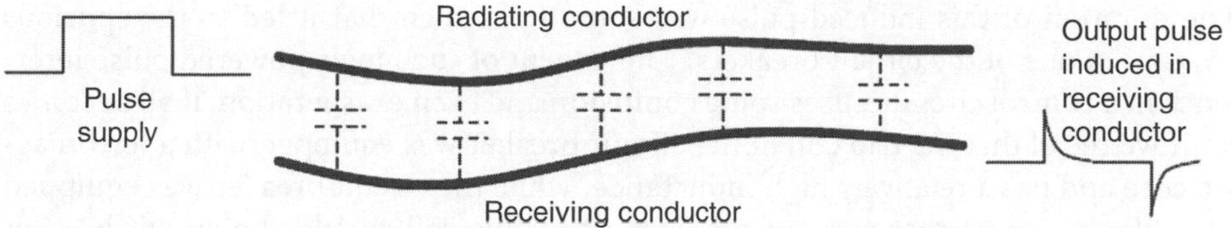

Fig. 4.35 Pulse interference moving between conductors via a capacitance coupling.

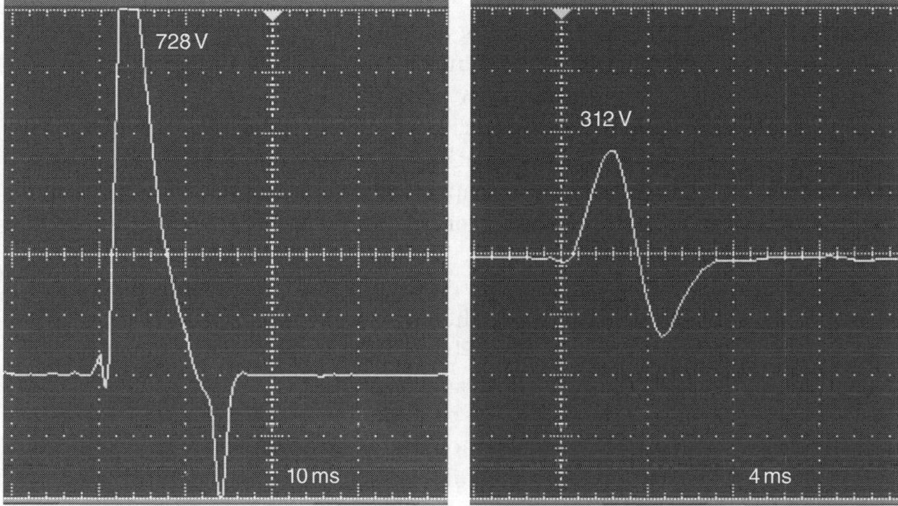

Fig. 4.36 One set of inducted pick-ups from one control cable over another: on the left is the unshielded cable; on the right is a cable shielded from both sides.

induced inside the shield is short circuited via the earth. When connecting the shield to the ground bus it is worth bearing in mind that no 'wire-wrapping' of the connecting cable is permitted, the same goes for twisting the long connecting cable between the shield and the ground bus into a ring. Each additional winding of this cable increases the impedance of the earthing system at high frequencies and dramatically reduces its effectiveness.

Sometimes the sources of high-power interference that affect substations are not clear and not self-evident. For example, at one of the Russian substations incidents were recorded of the activation of one of the high-voltage circuit breakers following a signal that was sent to the trip coil of another circuit breaker. The control cables leading to the trip coils were not shielded and ran in a communal cable tray for a distance of around 25 m. Experiments using voltage oscillation conducted at this substation, see Fig. 4.36, demonstrated that pulses with an amplitude of between 500–728 V could enter the coil in the circuit breaker that had been activated when a control voltage of 220 V was fed to the coil in the second circuit breaker.

The duration of this induced pulse was sometimes such that it led to the spurious activation of one of the circuit breakers. The advent of such high-powered pulse interference in a control circuit causes some confusion and even exasperation. It all becomes clear if we recall that the trip coil in the circuit breaker was equipped with a ferromagnetic core and has a relatively high inductance, while the circuit breaker was equipped with a block-contact that cuts off the current in this coil should the circuit breaker activate. As is well known the energy dissipated during an outage of an inductive current can be very significant. Following shielding of the control cable of one of the circuit breakers from both sides the strength of the inductive pulse interference on the second cable was greatly reduced, see Fig. 4.36, and the incidents of activation of the second circuit breaker disappeared completely.

The problem with the double sided earthing of the shield could arise simply out of the steady flow of considerable alternating currents through the central conductor (these are typically industrial frequency currents), which cause significant inductive currents within the shield, leading to it heating up badly. As a result, a cable with a larger cross section could be used (to reduce the overheating in the cable insulation) or one of the ends of the shield needs to be earthed via a capacitor. A capacitor would demonstrate a larger resistance for industrial frequency currents and a much lower resistance for high-frequency interference.

In some cases, a situation could arise where a significant interference pulse current could flow through a shield that has been earthed from both sides, which would give rise to interference in the central core. This could happen for example given the impact of a current from a bolt of lightning flowing in the cables that are located in close proximity to the control cables for the earthing elements or under the influence of a current from a nearby short circuit, see Fig. 4.37. As was demonstrated in [4.18] if a current from a bolt of lightning of 100 kA were to flow in the earthing electrode, even if the shield cable were earthed from both sides, the maximum value for the interference in the cable's central core could reach 8.2 kV, which far outstrips the resilience level of a DPR.

In these instances, it is not necessary to either change the route of the control cables (moving them away from high-power switching apparatus, lightning conductors,

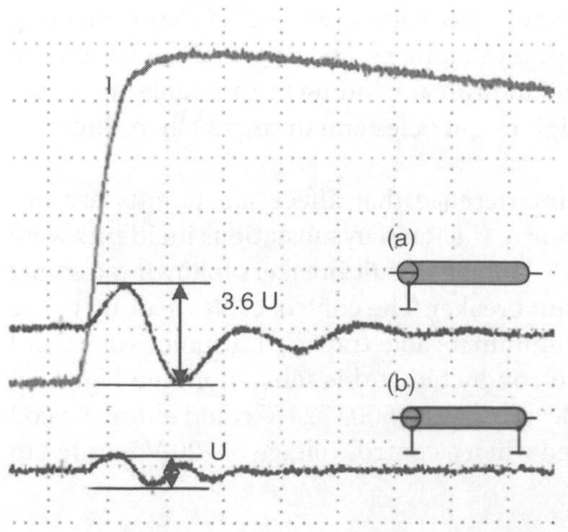

Fig. 4.37 Voltage interference on control cables with one sided (a) and two-sided (b) earthing as a pulse current (I) flows through the earthing electrode.

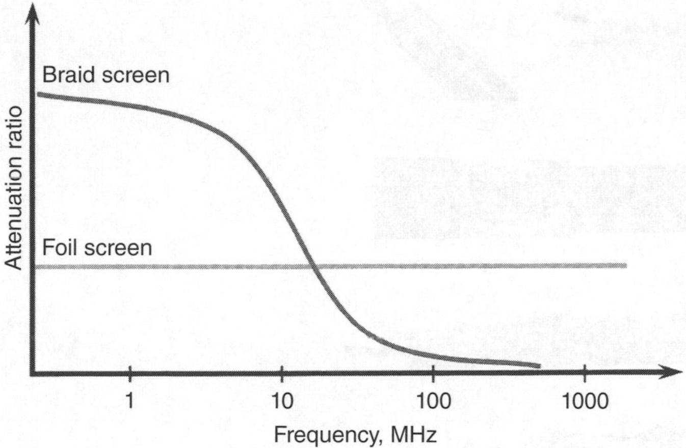

Fig. 4.38 The dependency of the shielding coefficient on the frequency for braiding and foil shields.

dischargers etc.) or reduce the voltage differences between the earthed ends of the shield cable when it comes under the influence of high-power pulse interference.

The latter is resolved by laying a copper bus alongside the cable that is earthed from both sides, and is known as a *potential equalizing bus*. The action of this bus is brought about by the fact that the impedance of a copper bus at high frequencies is considerably less than the impedance of the earth (or even that of the shield) and as such the greater part of the high-frequency interference pulse current would flow through this bus, and not through the shield. Naturally, these measures would be most effective if they were applied at the design and manufacturing stage of building a new substation and not as a 'band aid approach' at an old substation.

Control cables of course should be shielded with twisted pairs. The lowest minimal requirement for a shield is a high-density braiding (not less than 85%). Cables with double braiding demonstrate the best shielding effect. At relatively low frequencies up to several dozen megahertz braiding demonstrates a better shielding than foil, for the most part due to its thickness. However, the shielding properties of braiding deteriorate rapidly and become inadequate up to a frequency of 100 MHz. At the same time foil has flat amplitude-frequency characteristics at high-frequencies, retaining its satisfactory shielding capabilities right up to dozens of gigahertz, see Fig. 4.38.

Therefore, the preference is for cables to be given a combination of multi-layered shield containing both braiding and foil, see Fig. 4.39.

It is obvious that special types of control cables need to be used in the new designs that combine a paired cable with foil shields for each of these pairs with a triple-layer combined overall shield such as a 48-core RE-2X(ST)2Y(Z)Y PIMF type cable, as in Fig. 4.40 for example.

This is an ideal scenario, but what can be done with the dozens of old control cables already wired into existing protective relay cabinets? Replace them with new ones? In many cases this is very complicated and time consuming. Fortunately, many companies (such as Holland Shielding Systems BV) produce special mesh tape, with which old, unshielded control cables can simply be bound, and a mesh sleeve, which can be wound onto unshielded, or poorly shielded cables, see Fig. 4.41.

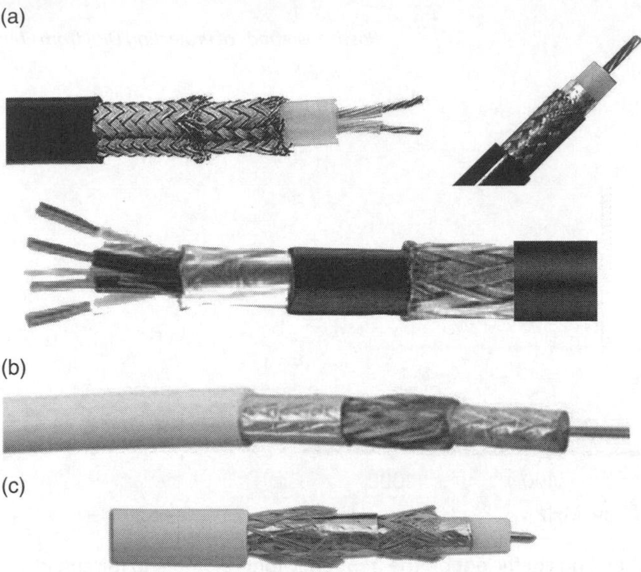

Fig. 4.39 Cables with a (a) double, (b) triple and (c) quadruple shielding combination (of braiding and foil).

Fig. 4.40 An RE-2X(ST)2Y(Z)Y PIMF type cable that is typically highly-resilient to interference (for the transfer of analogue and digital signals of up to 200 kbit/s; the cable uses a paired structure with a shield made from polyester foil for each pair; the triple-layer overall shield is made from foil and steel wire; the external insulation is made from cross-linked polyethylene (XLPE); the cable contains up to 24 pairs of conductors; it can be laid in the open or underground; it demonstrates excellent mechanical durability).

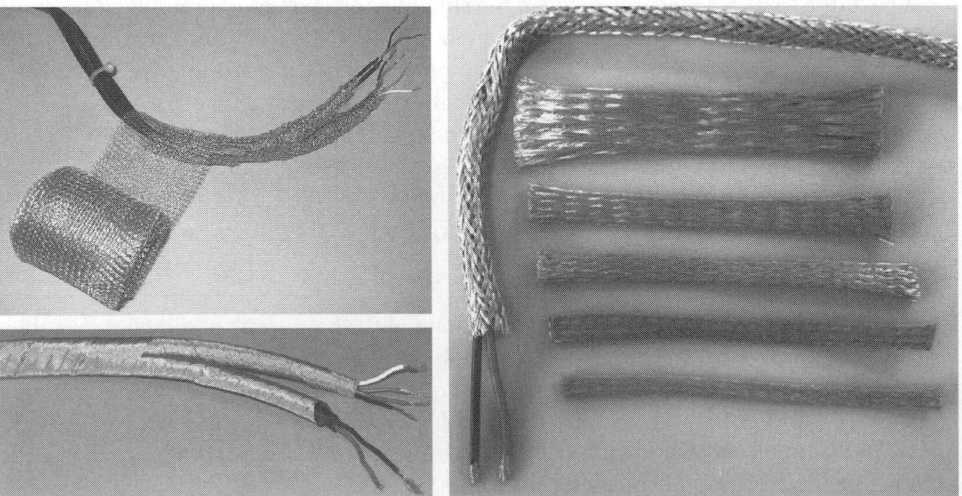

Fig. 4.41 Materials for self-made shielding of unshielded cables.

4.6 Design Changes to DPR

4.6.1 Analogue Input Points

Analogue inputs are elements that link a DPR with external current and voltage circuits by means of instrument transformers (CT and VT), therefore these elements would take the impact of HEMP in the first instance.

In a DPR input, CTs are the simplest in terms of design. As a rule, they represent a multi-loop secondary coil, wound onto a ferromagnetic core and a primary coil, consisting of several windings of thick insulated cable wound on top of an insulated secondary coil, see Fig. 4.42. Methods by which the resilience of this design to the threat of high-power pulse voltages can be enhanced are sufficiently simple and consist of the following:

- Encapsulating the secondary coil by immersing it in an epoxy compound and curing it in a vacuum, see Fig. 4.33.
- Using a high-voltage insulation cable to manufacture the primary coil;
- Using additional shields and semiconducting coatings, that equalize the electrical field in this design of CT;
- Using a magnetic core with an insulating surface.

Dozens of different types of flexible cables in a high-voltage insulation made from silicon, polyethylene, Teflon designed for 10–25 kV are produced by a number of companies: Teledyne Reynolds, Multi-contact; Allied Wire & Cable; Wiremax; Dielectric Sciences Inc., Axon' Cable, Dalburn Electronics & Cable, Sumitomo Electric, Belden, The Experimental Design Bureau of the Cable Industry; the Joint Stock Company 'Redkiy Kabel' and many others.

Recommendations to enhance the resilience of VTs are analogous, with the exception that instead of a flexible cable with high-voltage insulation as a primary coil a winding wire is used with enhanced third class insulation in accordance with the IEC 60317–0-1 'Specification for particular types of winding wires – Part 0–1: General requirements – Enamelled round copper wire made from Polyimide', and both coils are epoxy encapsulated in a vacuum, see Fig. 4.43. Since the broadening of the cross section of the

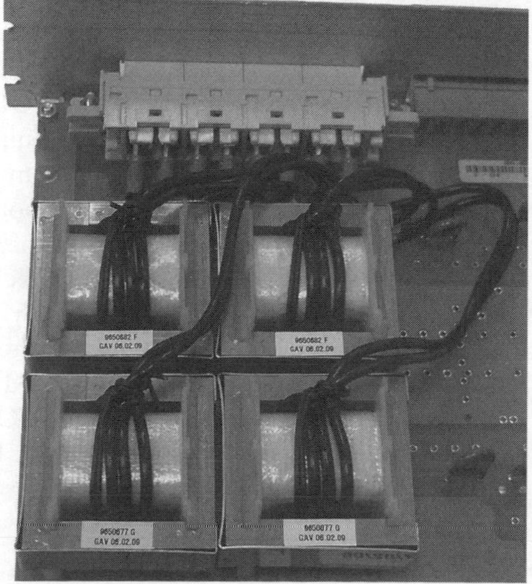

Fig. 4.42 A fragment of an analogue input module for a DPR with a built in CT. The primary coil, consisting of four windings of flexible, insulated black cable can be clearly seen.

Fig. 4.43 CTs manufactured according to a capsule design with a secondary coil, installed in a plastic casing and immersed in an epoxy compound that is cured in a vacuum. The primary coil, consisting of a single winding of flexible insulating cable can clearly be seen.

winding wire is accompanied by an automatic increase in the thickness of the insulation and in its electrical resilience then it follows that there is a striving to use a wire with a thicker cross-section, despite the natural increase in the dimensions of a VT. Several manufacturers produce winding wires with insulation made of polyimide, which is able to withstand one and a half times the voltage or even double the voltage in comparison with the standardized version in accordance with the IEC 60317-0-1 standard for example, produced by the English company P.A.R. Insulations & Wires Ltd, or the Turkish company Bemka A.S. among others.

4.6.2 Discrete Input Points

The insulation of discreet (logic) inputs across practically all types of DPR is provided by opto-isolators (optocouplers). As a rule, these are miniature optocouplers in standard DIP-4, DIP-6, DIP-8, and SOP-4 casings. Electrical resilience of insulation between the internal photo-emitting and photo-detecting equipment in these optocouplers can reach up to 5–7 kV r.m.s. In reality, however, these optocouplers, which are installed on a printed circuit board cannot withstand these voltages due to a breakdown between the pins across the surface of the printed board. At the same time optocouplers accommodated in special casings and with input and output pins spaced out around the casing are widely represented on the market, see Fig. 4.44. They can withstand a voltage between the input and output pins of up to 12–25 kV. These are OC100 optocouplers produced by Voltage Multipliers Inc.; HV801 optocouplers produced by Amptec Inc.; OPI1268S produced by TT Electronics; 5253003120 produced by Standex Meder Electronics among others. It is these same optocouplers that need to be used in discrete input on DPR in order to enhance their resilience to HEMP.

Fig. 4.44 An external view of different types of optocouplers with an insulating input/output voltage up to 12–25 kV.

The layout of DPR is such that the first elements to which a signal entering a discrete input point would be applied are the varistors that protect the optocouplers from overvoltages. Next would be the damping high impedance resistors, which would reduce the level of the input voltage (this is typically a 230 V direct current) down to the operating voltage for the optocouplers input circuit, which means that the current in this circuit does not exceed several milliamps. By using the TVS-diodes (see previously) in place of varistors the discrete inputs end up well protected not only from switching overvoltages, compared to when varistors are used, but also from short, high-power pulse voltages from the E1 HEMP component, should they get through to these inputs. Very fast acting modern optocouplers, particularly those based on modern photodiodes, which can reach 10^{-9} seconds is another problem. Therefore, with the aim of protecting the resilience of optocouplers to interference a different form of protection is required to protect them from the threat from a short E1 pulse, which could be achieved by shunting the optocouplers input using a high-frequency ceramic capacitor, thereby reducing the speed of reaction of the optocouplers and thus enhancing its resilience to interference.

4.6.3 Output Relays

Use of output relays in DPR with an enhanced electrical insulation resilience are one of the measures by which the resilience of a DPR to the threat from HEMP can be enhanced. The use of reed switch relays is very promising, these are made from new,

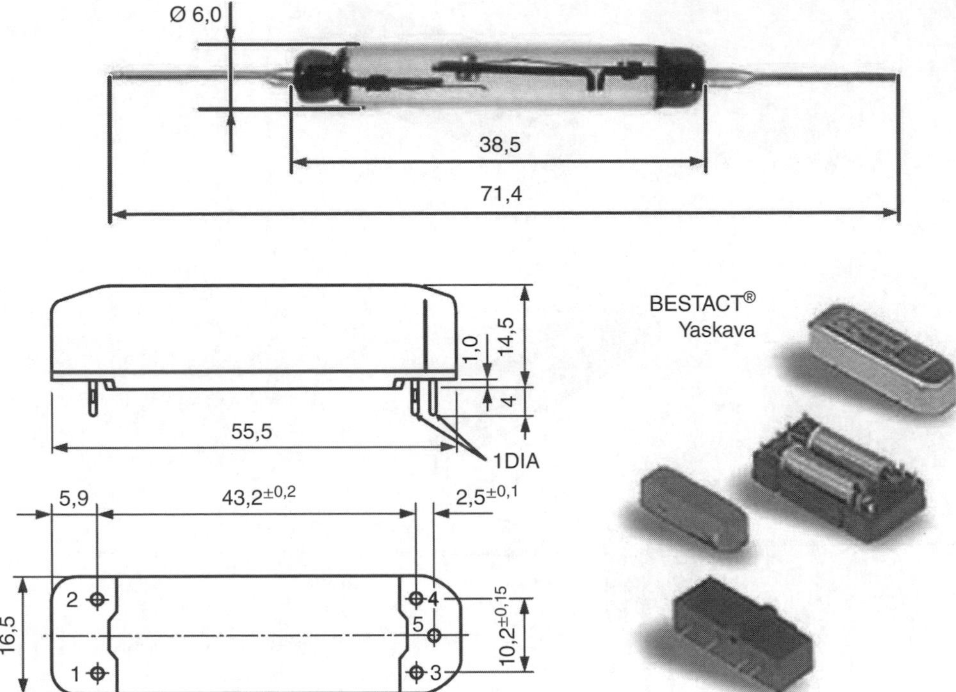

Fig. 4.45 A high-power R14U (R15U) type reed switches with a dual contact and a relay at its base, this example is produced by Yaskawa.

small, R14U and R15U high-power reed switches, with dual stage switching and are produced by Yaskawa under the trademark BESTACT®, see Fig. 4.45. These types of reed switches have a dual contact (a main one and an arc-suppressing one) with a subsequent switching, which enables an active inductive loading to be activated with a current of 15A and a voltage of 220V DC and 30A with a voltage of 220V alternating current. Based on these reed switches the company produces relays of different types, such as the R1-B14T2U for example. One of the distinguishing features of the reed switch relay compared to other types of electromagnetic relays lies in the simplicity of its design (a reed switch and a coil) and the potential to ensure a very high level of insulation (dozens of kilovolts) between the reed switch and the coil by using simple technology. This potential in a reed switch relay is very important in terms of its use as an output relay for a DPR that is protected from HEMP, and it could be realized on the basis of the author's existing designs, which are described in [4.19].

4.6.4 Printed Boards

The resilience of modern printed boards to pulse voltages with surface mounting elements is not just dependent on selecting the right electronic components but also on the breakdown voltage between the pins and the distance between the printed conductors (which are very small owing to the high density of their installation). Therefore, one of the additional means of enhancing the resilience of a DPR given the threat from HEMP could be to coat the board entirely on both sides with a special high-voltage lacquer. An example of this lacquer could be the products manufactured by Vol Roll under the trade mark Damicoat® – types 2405-01, 2407-01 and so on. These lacquers have an electrical insulation resilience of 70–100 kV/mm. Since printed boards coated in this lacquer become completely unserviceable, this gives rise to additional requirements for the design of DPR: the number of printed boards that form DPR should be increased so that should one of the function modules fail just this one element should be replaced and not a whole group of function modules located on a shared printed board. That is to say each function module (the power supply module, discrete inputs module, analogue input module, central processor unit (CPU), output relay module) should be housed on a separate extending printed board that would be linked to other printed boards by means of a coupling via the cross board.

Not only is this approach necessary in connection with the unserviceability of individual DPR modules but is very desirable in as much as it assists in solving the problem of the standardization of DPR modules and their universalization [4.20]. Another advantage of this design of DPR consisting of separate unserviceable functional modules is the potential for using a new (as far as relay protection is concerned) criteria for assessing reliability in place of the so-called, to put it mildly, strange criteria such as the 'mean time to failure' with its fantasy figures over the last 50–90 years that do not bear any relation to the real (and not fictitious) reliability. This criterion is known as the 'gamma percentile time to failure' and is characterized by a running time in accordance with which the failure of a piece of equipment does not occur in line with a specific reliability figure expressed as a percentage. For example, a 95% time to failure over the course of not less than 5 years means that no more than 5% of the devices in operation should fail over the course of 5 years. Using this clear and convenient indicator the consumer can monitor the number of modules that have failed over a specific period of time and can submit a complaint to the manufacturer if in this period of time a significantly greater number of modules have failed than was guaranteed by the manufacturer. Using this indicator, the consumer would be less drawn to the future market for universal modules [4.20] and would select the most suitable variant on the basis of price/quality. In addition to this the manufacturers would need to provide an indication in the technical and tender documentation for the average service life of individual modules, as well as recommendations for the periodicity of the preventative replacement of these modules with the aim of maintaining a high level of reliability in relay protection. For example, for a power supply module this could be 8–10 years; for a logic input module – 12–15 years; for a central processor module – 15 years; for an analogue input module – 17–20 years and so on. This data should be familiar to a conscientious manufacturer that has been monitoring the statistics for failures and damage among their products.

4.7 Construction Materials

One type of protection is to protect buildings and individual premises from penetration by EMP. The most powerful protection lies in the use of special panels, which have layers that contain, reflect and absorb electromagnetic interferences, see Fig. 4.46.

However, to create a fully shielded premise is a sufficiently expensive undertaking. Therefore, the cheaper interim variants using, for example, protective paints, films,

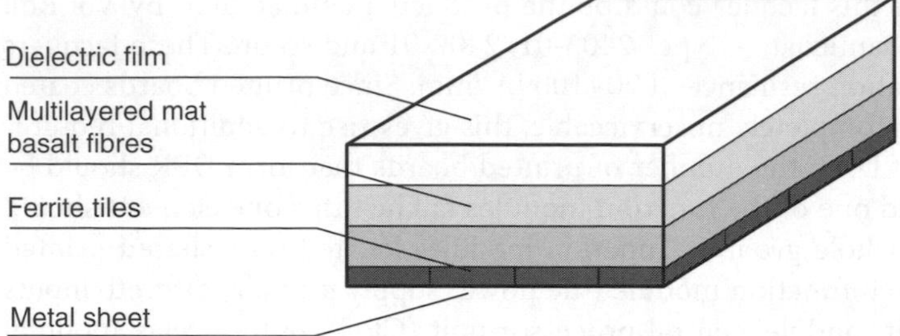

Fig. 4.46 The structure of a 'ferrilar-5' integrated panel used to protect premises.

curtains, drapery and so on are of particular interest. In the last few years, significant results have been achieved in the field of creating electrically conductive paints and coatings, and also construction materials that have unique properties and can be used widely, as well as transparent, electrically conductive coatings that can be applied to glass.

Electrically conductive paints, coatings and sprays based on copper, aluminium, bronze, nickel and graphite powder are produced by a number of companies such as Caswell YSHIELD EMR-protection Company, Less EMF Inc., Gold Touch, Inc., Spraylat Corp., Cybershield, Applied Coating Technologies Ltd, and BM Industria Bergamasca Mobili S.p.A. A protective paint known as 'Tikolak' is produced by the Moscow based company Tiko. Tikolak is a new, universal non-metallic electrically conductive paint and lacquer material which is protected by a patent in the Russian Federation and which represents a filler mixture containing carbon and a polymeric bonding agent working together: an epoxy bonding agent 8–20%, a graphite and carbon bonding agent in a mass ratio of 0.1:1.0:11–39%, condensing agent 0.5–1.5%, and the rest is taken up by an organic solvent. According to Tiko data Tikolak is a capable of shielding electromagnetic emissions across a broad frequency range right up to 300 GHz. Tikolak, when it is applied to the internal or external surfaces of a building reduces the penetrative capabilities of electromagnetic emissions drastically (according to the manufacturer's data a single coat of Tikolak with a thickness of just 70 μm reduces the intensity of EMP by 3–3.5 times). Tikolak can be applied to different construction materials – graded density chipboard, wood, plywood, gypsum board or onto any flexible material, textiles, leather, films, paper and so on. Any kind of decorative material can be applied to a Tikolak coating – wallpaper, paint, ceramic boards and so on, and as such Tikolak costs much less than foreign analogues (around $70 for 1 kg).

Semiconductor films made from oxides produced from a range of metals are used to create a transparent electrically conductive glass: tin, indium and zinc among others. The technology used to create these types of glass is very complex, labour intensive and requires expensive equipment and qualified personnel. Tiko, mentioned previously, have developed and patented (under RF patent No. 2112076) a highly technological and economic method of applying electrically conductive coatings onto glass using indium and tin oxides. Transparent, electrically conductive glass is manufactured by a range of companies, such as Tycon Technoglass, Pilkington, Shenzen Wanyelong Industry Co. Ltd, InkTec and so on.

Alfapol, based in St. Petersburg, have created construction materials using shungite rocks, which combine the properties of standard construction materials with a sufficiently high electrical conductivity. This defines the capability of the material to provide shielding from electromagnetic emissions. According to Alfapol data shungite composite radio-shielding materials can be divided into two categories in terms of the capability of the shields made from these materials:

- Construction materials, which include concrete, brick, and mortar. These materials are capable of ensuring an attenuation of electromagnetic energy across a frequency range of more than 100 MHz to a level of not less than 100 dB. In terms of their physical and mechanical characteristics shungite based construction materials do not concede to their traditional analogues in construction.
- Shungite materials have undergone testing in construction (concrete in floor panels, brickwork) and have been recognized as complaint with all existing requirements.
- Renovation materials, such as finishing mortar and mastics, which can be used to convert standard structures to shielded ones. Mastics are capable of providing a screening effect to a level of not less than 30 dB across a range of more than 30 MHz at a layer thickness of 2–3 cm. The 'Alfapol ShT-1' finishing plaster at a layer thickness for the plaster of 15 mm across a frequency range from 10 kHz to 35 GHz provides a 10–15 dB attenuating effect.

Shungite is a group of solid, carbon mineral substances, which for the most part represent amorphic variations of carbon, and in terms of composition are similar to graphite. The chemical composition of shungite is changeable: on average it consists of 60–70% carbon and 30–40% ash. The ash consists of: 35–50% silicon oxide, 10–25% aluminium oxide, 4–6% potassium oxide, 1–5% sodium oxide and 1–4% titanium oxide, as well as a mixture of other elements.

As an addition to the walls of a premise containing shungite electrically conductive curtains or textiles can be added that have been impregnated with a coating manufactured by a number of companies, see Fig. 4.47.

Fig. 4.47 Electrically conductive films, threads, and textiles that are capable of attenuating electromagnetic interferences (down to 80 dB).

References

4.1 Whitaker J.C. *Electronic Systems Maintenance Handbook*, 2nd Edn - CRC Press (Taylor & Francis Group), Boca Raton – New York – London, 2001, 624 p.

4.2 Ilyin V.F., Ilyin N.V. Earthing in DPR cabinets - *Relay Protection and Automation*, 2015, No. 1. pp. 26–30.

4.3 Gurevich V.I. *The Vulnerability of DPR - Relay Protection. Problems and Solutions.* M.: Infra-Inzheneriya, 2014, 256 pp.

4.4 Grounding and Bonding in Command, Control, Communications, Computer, Intelligence, Surveillance and Reconnaissance (C41SR) Facilities. Technical Manual TM 5–690. Headquarters Department of the Army, 17 February 2002.

4.5 IEC 60364–5-548: 1999. Electrical installations of buildings - Part 5: Selections and erection of electrical equipment - Section 548: Farthing arrangements and equiponential bounding for information technology installations.

4.6 MIL-STD-188–125–1 High-Altitude Electromagnetic Pulse (HEMP) Protection for Ground-Based C4I Facilities Performing Critical, Time-Urgent Missions; Part 1: Fixed Facilities.

4.7 Nalhorczyk A.J., HEMP Filter Design to Meet MIL-STD-188–125 PCI Test Requirements. – *IEEE 10th International Conference on Electromagnetic Interference & Compatibility*, 26–27 Nov., 2008, pp. 205–209.

4.8 MIL-STD-461F Requirements for the Control of Electromagnetic Interference Characteristics of Subsystems and Equipment, 2007.

4.9 IEC 61000–4-4 Electromagnetic compatibility (EMC) - Part 4–4: Testing and measurement techniques - *Electrical Fast Transient/Burst Immunity Test*, 2012.

4.10 Application Notes Cat. 1: HEMP Filter Maintenance and Monitoring. Rev. 1. MPE Ltd., December, 2012.

4.11 Surge Protective Device Response Time, Application Note 9910–0003A, Schneider Electric, August 2011.

4.12 Power Quality Surge Protective Devices (SPD), Application Notes: Response Time ratings, DET-733 (8/10), General Electric.

4.13 Surface Mount Power TVS Diodes Deliver Optimal Protection for Power Supply. Application Note, Bourns, Inc., 7/14.e/ESD1435.

4.14 Goldman S.J., Selecting protection devices: TVS diodes vs. metal-oxide varistors - *Power Electronics*, June 1, 2010.

4.15 IEC 61000–4-25:2001 Electromagnetic compatibility (EMC) - Part 4–25: Testing and measurement techniques - HEMP immunity test methods for equipment and systems

4.16 Gurevich V.I. Problems with testing DPR for resilience to HPEM - *Components and Technology*, 2014. No. 12, pp. 161–168.

4.17 MIL-STD-220B Method of Insertion Loss Measurement, Department of Defense, 1959.

4.18 Kuznetsov M.V., Kungurov D.A., Matveyev, M.V. Tarasov V.N. Problems with the protection of input circuits in relay protection apparatus from high-power pulse overvoltages – *Relay protection and Substation Automation of Modern EHV Power Systems* (Moscow - Cheboksary, September 10–12, 2007).

4.19 Gurevich V. *Protection Devices and Systems for High-Voltage Applications.* Marcel Dekker, New York, 2003, 292 p.

4.20 Gurevich V.I. *Problems of Standardisation of Relay Protection. SPB.* The DEAN Publishing House, 2015, 168 pp.

5

Active Methods and Techniques of Protecting DPR from EMP

5.1 A New Principle in the Active Protection of DPR

There is a critical vulnerability in the structure of a contemporary DPR energy system [5.1, 5.2] in that, on the one hand, they are the most vulnerable to Intentional Destructive Remote Threats (IDRT) and, on the other hand, they are directly connected to a high-power circuit breaker, which affects the status of the energy system. Therefore, IDRT in the form of cyber-attacks along with HPEM are aimed first and foremost at the DPR [5.3–5.5].

An awareness of the problem of cyber security in DPR in recent years has led to the intensification of much research work, which for the most part has been aimed at upgrading the communication protocols designed for relay protection and at enhancing their cryptosecurity. Up until recently, all the efforts of specialists were focused on this line of work. As far as HPEM is concerned it was demonstrated in Chapter 2 that nobody is looking into this problem *seriously*. Meanwhile back 18 years ago when the problem of DPR had only just began to surface the author proposed an idea in general terms for the highly-effective combined protection of DPR both from cyber-attacks and from HPEM by using hardware generated and not software generated methods. This was a protection device that operated on the principle of shunting the sensitive DPR inputs by using fast acting electro-mechanical reed relays [5.6]. Subsequently the idea of using fast acting electromechanical reed relays jointly with DPR to reduce their vulnerability to IDRT was developed by the author a little more diligently [5.7, 5.8].

As we have already demonstrated more than once, *it is impossible to solve the challenge of enhancing the reliability of DPR given the contemporary functions of a DPR*, namely *functions that do not bear any relation to relay protection*. This refers to popular functions such as monitoring the serviceability of electrical equipment, remote control of circuit breakers and so on, while DPR should be used exclusively for the purposes of relay protection. Furthermore, there are a huge number of specialized devices on the market today that are designed specifically for these tasks like monitoring electrical equipment, from the simplest relays controlling the integrity of a circuit breaker coil circuit to the most intricate complexes controlling the composition of gases that are formed in transformer oils, or the level of partial discharges in insulation in real time. As far as remote control of circuit breakers using DPR is concerned if they are used in this way it is very difficult to distinguish sanctioned remote access from

Protection of Substation Critical Equipment Against Intentional Electromagnetic Threats,
First Edition. Vladimir Gurevich.

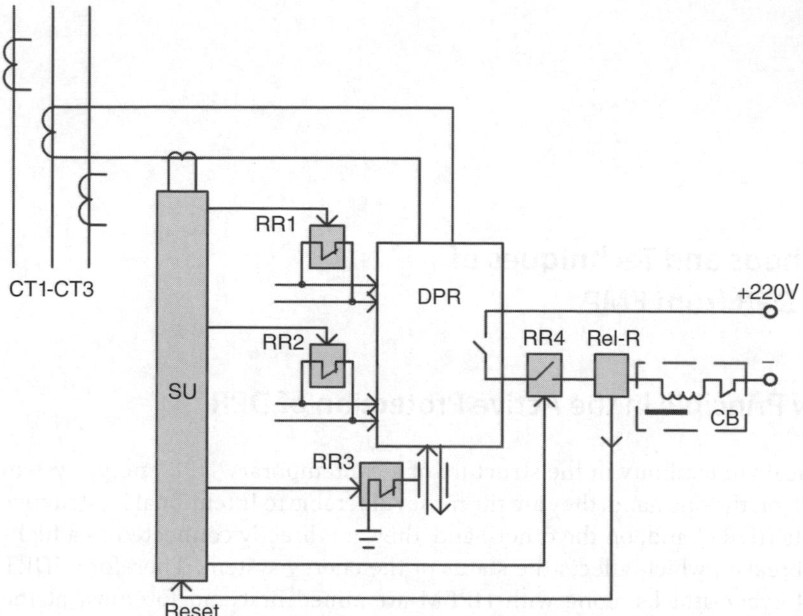

Fig. 5.1 A structural flow chart for a protection device designed to protect DPR from IDRTs.

spurious remote access, so therefore this function of a DPR should be banned. Moreover, if the functions are delegated, then protection from IDRT and from the remote control circuit breaker system can be provided using sufficiently simple hardware based approaches (see next).

The general idea that lies behind the principle of this proposed hardware based method of protecting DPR from IDRTs lies in the use, jointly with a DPR of a reed switch based electromechanical starting element (SE), which is functionally wired in series with a DPR and fast acting electromechanical actuators (RR-1 to RR-7), ensuring that the sensitive inputs to a DPR are blocked and the output circuit is disconnected, see Fig 5.1. The return of the active SE to its default condition is achieved as the circuit breaker is activated and is duplicated by a signal RESET that the short, pre-set time period has expired. Without either the current and/or the voltage activating this starting element the DPR cannot have any bearing on the mode of operation of the energy system, even if it is subject to the impact of IDRT, or simply to high-power electromagnetic interference. If the SE was activated and the DPR was unblocked, then there is nothing to stop the specific functions and the functional capabilities of a DPR being used. Furthermore, any false activation of this same SE would not have any bearing on the operation of a protective relay and as such no specific requirements concerning the validity of an activation of a starting element are required. The only important thing is that SE always activates prior to the DPR, that is to say that it should have a slightly reduced activation setting for this control parameter. If the activation of a starting element were to prove false and the DPR was not activated, then the device would automatically reset to its default condition. The principle technical requirements for this device are its excellent reliability, insensitivity to short pulses (across the micro and

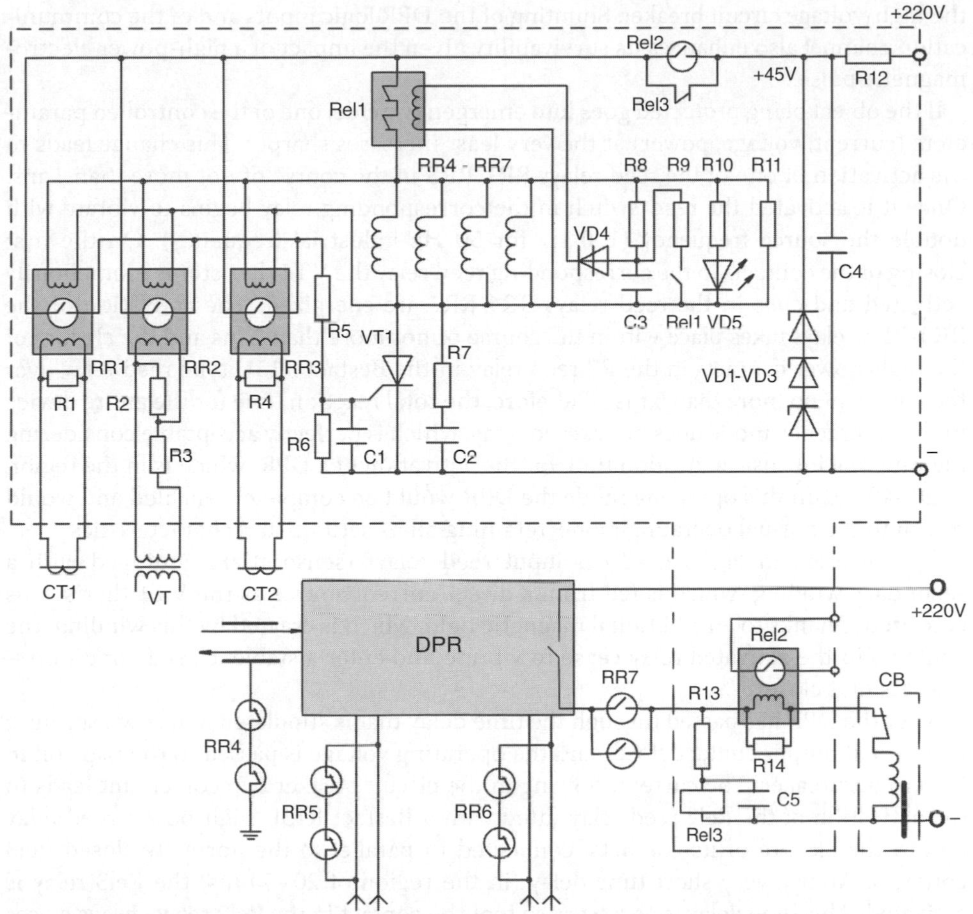

Fig. 5.2 A modified wiring diagram for a protection device designed to protect a DPR from IDRT.

nanosecond range) and to high-frequency interference, their resilience to significant overvoltages, a high level of galvanic separation from external circuits and a high speed of response to activation (several milliseconds). The reed switches are the elements that allow these challenges to be solved by simple technical means.

This chapter sets out a description of an upgraded device, designed to protect a DPR from IDRT and satisfies the requirements formulated here, see Fig. 5.2.

This device works as follows. In the default, when the device being protected is in its normal operating mode all the input reed relays (such as the current and voltage sensors etc.) RR1-RR3 are in the released position.

The VT1 thyristor is locked in, and the coils for the active reed-switch based relays RR4-RR7 are de-energized. The changeover RR5 and RR6 contacts disconnect and de-energize the DPR logic inputs, the RR4 contacts – the communication channel and the RR7 contacts break the DPR output circuit. Thus in this condition the DPR is completely disabled both in terms of the outputs and inputs and no IDRT or cyber-attacks can cause it to activate by accident or cause any unsanctioned closure of the trip coil in the high-voltage circuit breaker. Shunting of the DPR logic inputs and of the communication channel also enhances its survivability given the impact of a high-power electromagnetic pulse.

If the object being protected goes into emergency mode, one of the controlled parameters (current, voltage, power) at the very least increases sharply. This change leads to the activation of one of the reed relays RR1-RR3 in the course of not more than 1 ms. Once it is activated the reed-switch in the corresponding relay begins to vibrate with double the source frequency (100 Hz for 50 Hz industrial frequency). On the first closing of the contacts in the corresponding reed relay the VT1 thyristor is momentarily activated and coils in the reed relays RR4-RR7 are energized. The activation of the RR4-RR6 relays takes place within the course of not more than 1 ms and the closure of the high-power contacts in the R7 reed relay in the Bestact R15U type reed relay over the course of no more than 5 ms. Therefore, the total reaction time for the entire device to the emergency mode does not exceed 6 ms, which is perfectly acceptable considering the minimal intrinsic activation time for the activation of a DPR, which is in the region of 20–40 ms. In this operating mode the DPR would be completely enabled and would return to the normal operating mode, retaining all its settings and characteristics.

As is evident in Fig. 5.2, all the input reed relays (sensors) are equipped with a secondary winding, which is fed from a direct current source as the VT1 thyristor is activated. Thanks to an additional magnetic field, which is created by this winding, the contacts in the activated relay cease to vibrate and enter a stable, closed condition – after its first closure.

After the DPR has passed through the time delay that is stipulated in its own settings, its internal output contact closes and the operating voltage is passed to the trip coil in the circuit breaker. The current, flowing in the circuit breaker trip coil circuit leads to the activation of the Rel2 reed relay fitted with a Bestact R15U high-power reed relay and to the closure of its contacts, connected in parallel to the normally closed Rel3 contacts. After a very short time delay (in the region of 20–50 ms) the Rel3 relay is activated. This time delay is necessary so that the contact in the Rel2 relay always closes prior to the release of the Rel3 contact.

At the end of the circuit breaker operating cycle the block-contact opens and the trip coil feed circuit is disrupted. Moreover, the Rel2 relay is released and its contact disrupts the anode circuit in the V1 thyristor, which is momentarily cut off. In doing so it de-energizes the windings in the RR4-RR7 relays and the direct current windings in the RR1-RR3 relays. The entire device quickly returns to its default conditions, and is ready for another operating cycle.

If the activation of the device turns out to be spurious and the DPR does not pass on the command to disconnect the circuit breaker, the feed circuit for the V1 thyristor is temporarily disconnected by the normally closed Rel1 contact relay, after the capacitor C3 has been charged via the resistor R8 and the dynistor VD4 has been triggered. The capacitance of this capacitor and the resistance in the resistor ensure a time delay lasting a few seconds, which exceeds the maximum permitted time necessary for the DPR to complete its full operating cycle, so as not to interrupt its operation should it prove necessary. The activation of the Rel1 relay is very short, since immediately after its activation and the opening of the normally closed contact in the thyristor circuit, its normally open contact closes and the C3 capacitor discharges via the low resistance resistor R9, which ensures it is fully discharged and returns to its default. Moreover, the

VD4 dynistor is locked out and the coil in the Rel1 relay is de-energized. This ensures the mandatory return of the device to its default, should the activation prove unnecessary. The R11 resistor is required in order to increase the current flowing through the VT1 high-power thyristor and to ensure that it maintains its conductive state. The VD5 LED acts as an indicator of the condition of the device.

The contacts in the external relays, which are designed to activate/deactivate the internal functions of the DPR and that define the condition of the protective relay, should be connected alongside the DPR protected logic inputs (those that have been shunted into a normal mode), which are illustrated in Fig. 5.2, as well as to the logic inputs (Inp 1, Inp. 2) in this protection device. Furthermore, with the aim of enhancing the system's protectability there needs to be no less than two signal inputs for one event, which should be fed from two different sources. Inside the protection device these signals are galvanically isolated from external circuits by the additional contact reed relays RR7 and RR8, which enables the logic function 'AND', resulting in the activation of the protection device and the release of all the logic signals so they can fulfil their required operations. Moreover, these two sources do not necessarily have to be in the form of discrete signals (relay contacts). One of them could be a discrete signal, while the other could be an analogue signal, in the form of a current or a voltage, arriving at a corresponding reed relay input, for which the external contacts have been connected in accordance with the 'AND' logic function. Furthermore, the output of the relay in the second (unused) logic input can be shunted by the bridge S.

The resilience of these additional logic inputs to short pulse interference even with a very high amplitude is provided by a corresponding level of insulation between the winding and the reed relays and also by the intrinsic time for the activation of the reed relay. The latter is not able to function in a time span of less than several milliseconds, and as such is a very natural and highly effective filter for high-frequency and pulse interference.

With the aim of ensuring the required resilience to interference to all kinds of transient processes in the operational direct current feed circuits these additional inputs should have an input resistance that is significantly less than the standard logic inputs for a DPR, which is achieved by shunting them using the R15 and R16 internal resistors.

Although the windings in the reed relay inputs are significantly more resilient to overvoltages, than the semiconductor devices, measures have been taken inside this device in order to provide additional protection from overvoltages that arise during HPEM, using the VDR1 and VDR2 varistors.

In design terms the protection device is manufactured inside a screened casing, which is analogous to DPR casings but with one exception – that it does not feature a screen, however, access to the regulation threshold pick-up for the SE reed relays is provided.

With the aim of enhancing the reliability of the device and its resilience to HPEM only a few semiconductor devices are used inside, and those that have been chosen have large reserve of current and voltage, which are not usually used in standard industrial equipment. Thus, at an operating voltage of 45 V the VT1 thyristor was chosen with a maximum working voltage of 1200 V, and at an operational current of fractions of amperes, it is capable of operating in currents of dozens of amperes and of passing short current pulses of hundreds of amperes. The VD1-VD3 Zener diodes, and the VD4 dynistor have also been chosen with a much greater reserve of output. The intermediate Rel1 and Rel3 relays have been chosen as they are sealed and have contacts with an

Table 5.1 The principle parameters for various different types of fast-acting high-power vacuum reed relays.

Parameter/Type of relay	MRA 5650G	KSK-1A75	HYR 2016	HYR 1559	MARR-5	KSK-1A85
Contact type	NO	NO	NO	NO	NO	NO
Switching voltage, V	1000	1000	1000	1500	1000	1000
Switching current, A	1	0.5	1	0.5	0.5	1
Switching power, W	100	10	25	10	10	100
Breakdown voltage, V	1500	1500	2500	1500	2000	4000
Activation time, ms	0.6	0.5	0.8	0.4	0.75	1.0
Release time, ms	0.05	0.1	0.3	0.2	0.3	0.1
Dimensions, mm	D = 2.75	D = 2.3	D = 2.6	D = 2.3	D = 2.66	D = 2.75
	L = 21	L = 14.2	L = 21	L = 14.2	L = 19.7	L = 21
Sensitivity, ampere-windings	20–60	15–40	15–70	15–50	17–38	20–60

enhanced capacity. The general recommendations for choosing the element base for the proposed active protection device are set out next.

Miniature vacuum reed switches, which are able to withstand a test voltage of not less than 1 kV and which have an intrinsic activation time of 1 ms, can be used as a sensitive threshold element in the SE, see Table. 5.1.

Since it takes no more than a few seconds typically for a DPR to function in emergency mode, a reset time of 5–10 s for the system to return to its standby mode is more than enough for a DPR to complete the normal operating cycle.

Since the total current consumed by the windings in the RR4-RR7 relays may end up less than the latch-up current (I_L) and the holding current (I_H), in the VT1 thyristor, the diagram in Fig. 5.2 is augmented by the high-power R11 resistor, which increases the total current flowing through the thyristor up to 250–300 mA. Although special thyristors are available on the market which have enhanced sensitivity and latch-up and holding currents that do not exceed 10 mA (TS820–600, TIC106, BT258–600R, X0402MF, MCR708A1 etc.) their use in this device is not recommended since this could reduce its resilience to interference.

In Table 5.2, the parameters for several different types of the most suitable thyristors for use in a starting element are set out. In order to enhance the resilience to interference in a starting element additional RC-elements have been used.

With the aim of acting as active relay contacts, which block a DPR output contact, gas filled Bestact R15U reed switches produced by Yaskawa can be used (see Fig. 5.3), these are designed to energize currents of up to 30 A at a voltage of 240 V in a timescale that does not exceed 5 ms.

High-voltage miniature vacuum changeover type reed switches of different types, which contain a normally closed contact part, Table 5.3 can be used to shunt sensitive (non-current) DPR inputs.

For electromechanic relays, the likelihood of 'false activation' type failures is incomparably less than the likelihood of 'non-activation'; therefore, their connection in parallel (as opposed to a simple parallel connection of a DPR) unequivocally enhances the reliability of the protective relay.

Table 5.2 The most important parameters for several different types of high-voltage thyristors, which are designed for soldering onto a printed circuit board.

Parameter / Type	CLA50E-1200HB	25TTS12	30TPS12 30TPS16	BTW68–1200
V_{RRM}/V_{DRM}, V	1200	1200	1200 1600	1200
$I_{T\ (RMS)}$, A	79	25	30	30
$I_{T\ (AV)}$, A	50	16	20	19
I_{TSM}, A	650	300	250	400
I_{GT}, mA	50	60	45	50
I_L, max., mA	125	200	200	40
I_H, max., mA	100	100	100	75
dv/dt, V/μs	1,000	500	500	250
T_{GT}, μs	2	0.9	0.9	100
T_J, °C	−40 + 150	40 + 125	−40 + 125	−40 + 125
Casing	TO-247	TO-220AC	TO-247AC	TOP3 ins.

Parameter / Type	CS 20–12io1 CS 20–14io1 CS 20–16io1	CS 30–12io1 CS 30–14io1 CS 30–16io1	CS 45–12io1 CS 45–16io1
V_{RRM}/V_{DRM}, V	1200 1400 1600	1200 1400 1600	1200 1600
$I_{T\ (RMS)}$, A	30	49	75
$I_{T\ (AV)}$, A	19	31	48
I_{TSM}, A	200	300	520
I_{GT}, mA	65	65	100
I_L, max., mA	150	150	150
I_H, max., mA	100	100	100
dv/dt, V/μs	1000	1000	1000
T_{GT}, μs	2	2	2
T_J, °C	−40 + 125	−40 + 125	−40 + 140
Casing	TO-247AD	TO-247AD	TO-247AD

For normally closed auxiliary contacts in DPR shunting inputs, an increase in reliability can be achieved by connecting these contacts together in series.

The capacitor in the time-delay RC-circuit, as with all the other elements of this device, has been chosen for its high quality and for its excellent reserve (up to five times) of operational voltage, see Table 5.4.

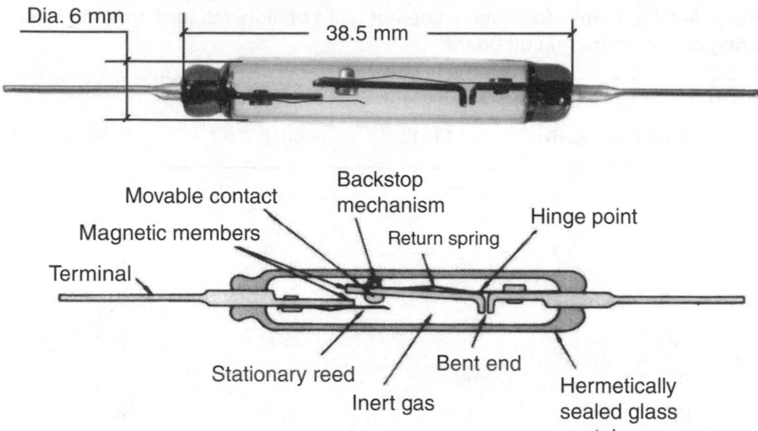

Fig. 5.3 A high-power gas filled Bestact R15U reed switch produced by Yaskawa with two-stage switching.

Table 5.3 The main parameters of different types of changeover reed switches.

Parameter / Type and manufacturer	GC 1917 Comus	HSR-803R Hermetic Switch	HSR-834 Hermetic Switch	HSR-V933W Hermetic Switch	DRR-DTH Hamlin
Max switching power, W	60	25	100	100	50
Max switching voltage, V	400	250	500	500	500
Max switching current, A	1	1	3	3	0.5
Breakdown voltage, V	1000	1000	1000	1500	1200
Activation time, ms	4.0	3.6	2.0	4.2	4.5
Release time, ms	0.15	4.2	1.0	3.7	7.0
Dimensions of the cylinder, mm	D = 5.6 L = 36	D = 5.3 L = 32	D = 5.3 L = 34	D = 5.3 L = 33	D = 5.5 L = 39.7

Table 5.4 The main parameters of several types of high quality capacitors used in time-delay RC-circuits.

Type of capacitor	Manufacturer	Capacitance and voltage	Dimensions, mm	Operating temperature, °C
B43504B2477M	EPCOS	470 μF, 250 V	Dia. 30 × 30	−40 + 105
B43505A2477M	EPCOS	470 μF, 250 V	Dia. 30 × 35	−40 + 105
EETHC2E471CA	Panasonic	470 μF, 250 V	Dia. 25 × 30	−40 + 105
MAL215933471E3	Vishay	470 μF, 250 V	Dia. 25 × 40	−25 + 105
MCHPR250V477M25X41	Multi-comp	470 μF, 250 V	Dia. 25 × 41	−25 + 105
381LQ471M250J022	Cornell Dublier	470 μF, 250 V	Dia. 25 × 30	−40 + 105

Table 5.5 The main parameters of several different types of Zener diodes with a power of 10–20 W and a nominal voltage of 15 V.

Param. / Type	NTE5191A	1N2979	BZY93-C15
P_D, W	10	10	20
V_Z, V	15	15	15
I_{ZM}, mA	560	560	1000
I_{ZT}	170	170	170
Z_{ZT}, Ω	3	3	1.2
I_R, μA	10	5	50
T_{OPR}, °C	−65 + 175	−65 + 175	−55 + 175
Casing type	DO-4	DO-4	DO-4

Three VD1–VD3 Zener diodes connected in series with a nominal voltage of 15 V and a power of 10 W each have been chosen to act as voltage divider.

Given a very small in house use of this design, the excellent reserve of output ensured that the Zener diodes did not overheat and enhanced their reliability, as did the capacity to absorb high-energy pulse overvoltages. The parameters of the most suitable Zener diodes to achieve this goal are set out in Table 5.5.

The following types of dynistors can be recommended to use as a VD4 (Fig. 5.2) with a barrier voltage of 24–36 V and a pass current of 1–2 A: NTE6407, DB3, BR100/03, CT-32, HT-32 amongst others. Sealed neutral electromagnetic relays (which are known as Full Size Crystal Can Relays in technical literature) with two switching contacts (two normally closed contacts are used to enhance reliability), and with a switching current of 2–5 A and 24 V DC winding are used as the relay Rel (see Fig. 7.2). The following series of relays can be cited as an example of these types of relays: REN33, REN34, REK134, RES48, 782XDXH, H872, B-7, FW, SF, G2A-434ADC24, HGPRM-B4C05ZC and 2B-7506, among others.

There is no specific need to tune the pickup threshold for this device. What is important is that it always activates prior to the DPR, given any suspicious mode in the control circuit, since any spurious activation of this device as a result of inaccurate tuning does not have any impact on the behaviour of the DPR that is under its protection.

The use in this proposed device of highly reliable components that have been selected for their excellent reserve of current, voltage, and capacity, which enables them to operate across a wide temperature range, the very limited number of these components, the high level of galvanic isolation, as well as the duplication of the most important elements provides excellent reliability for a DPR and given the impact of high-power electromagnetic interference, cyber-attacks and HPEM, one which corresponds to the reliability and resilience of old electromechanical relays.

All the elements in this protection device should be accommodated in a separate casing and fitted with plug and socket units or connecting blocks to enable them to connect to a DPR. DPRs that are protected using this device can be connected (where necessary) to work in parallel for the very important power installations. If this device is used it is also possible to connect additional electro-mechanical protective relays in parallel with DPRs with a time delay of 0.1 s [5.7].

Evidently, specific circuitry solutions may differ from those described in this chapter, however the proposed approach to solving this problem would undoubtedly encourage an improvement in the reliability of DPR. There is no doubt that specific operating modes for protective relays will be found as well as ways in which individual protective relays can work together, which would be problematic for this proposed method of protection. This is completely natural and is to be expected. These specific circuitry solutions may require either an overhaul of the proposed method, or a change in the pattern of cooperation between relays. The description of this hardware based approach is intended only as a confirmation of the potential for a technical realization of the concept of protecting DPR from IDRT using a hardware based rather than a software based approach, and could serve as a starting point for specific hardware manufacturers producing equipment that is suitable for industrial production. Subsequent efforts should be focused on developing designs for input elements (current and voltage sensors) based on reed switches with a regulated pickup threshold.

5.2 Current and Voltage Sensors with Regulated Pickup Threshold based on Reed Switches

Reed switch relays are widely used in technology and are produced by a wide range of companies. The advantages of reed switch relays, such as the fact that they are air tight, have a long service life, their speed of pickup, the special gaseous environment or vacuum that houses the contact components and the fact that there is no need to control and clean the contacts, the high level of galvanic isolation between the input (the control coil) and the output (reed switch), as well as their concise and stable pickup threshold, make them indispensable in a whole host of automation and measuring technology. However, not one of the reed relays produced in industry possess the ability to regulate the pickup threshold, which the design of these relays would require.

We will examine the most acceptable designs of relay for this purpose that have a regulated pickup threshold, see Fig. 5.4.

The simplest variant is the design set out in Fig. 5.4. with a coaxial reed switch and a control coil displacement, and with a movement of the reed switch within the coil shaft. This relay demonstrates the maximum sensitivity when the clearance between the contacts in the reed switch is located in the centre of the coil.

In moving this clearance from the centre of the coil the sensitivity of the reed switch to the current flowing through the coil is reduced. However, the practical realization of this design would appear not to be so simple, see Fig. 5.5.

In order that the reed switch move the assembly, which is similar to a worm gear reducer, in which the shaft 1, with an internal thread rotating about its own axis makes the shaft with an external thread, which is located at the end of the component 2 with a reed switch pressed into it, rotate. Aside from the complexity, one disadvantage of the design is the considerable length of the relay L, which exceeds a length, triple that of the reed switch tube. Another disadvantage of the design is that the reed switch leaves the effective magnetic shielding zone as it is pulled out of the coil. The electrical resilience of the insulation between the coil and the relay in this design does not exceed 1 kV.

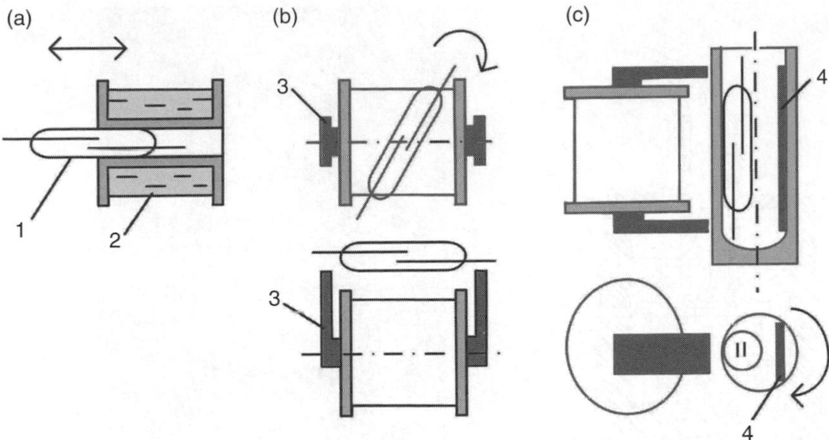

Fig. 5.4 The structural layout of reed switch relays with regulated pickup threshold: a – with a coaxial reed switch movement within the coil; b – With the reed switch shaft rotating about the coil shaft and with the reed switch carried externally; c – with an eccentric reed switch movement and a magnetic shunt. 1 – reed switch; 2 – Coil with a winding; 3 – Ferromagnetic core; 4 – Ferromagnetic plate shielding (magnetic shunt).

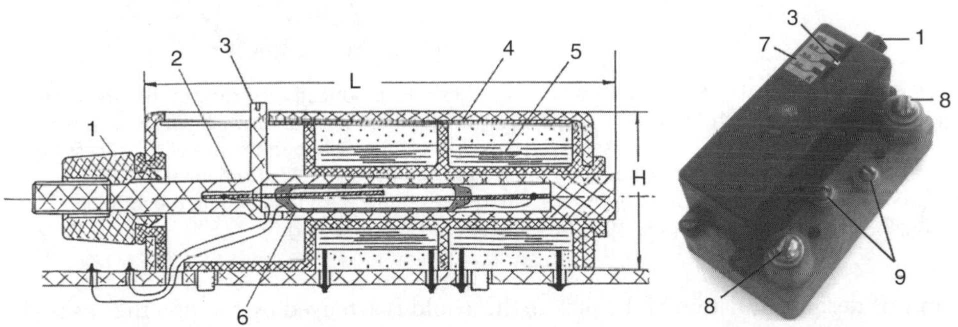

Fig. 5.5 A reed switch relay with a regulated pickup threshold, an axial reed and coil displacement, and an axial movement of the reed switch. 1 – The rotating tuning handle; 2 – A mobile, plastic component with a reed switch pressed into it; 3 – Reed switch position indicator; 4 – Ferromagnetic screen; 5 – A coil with a winding; 6 – Reed switch; 7 – Scale; 8 – Winding outputs; 9 – Reed switch outputs.

Another design, which is easier to produce is that proposed in Fig. 5.6; in this design a ferromagnetic core is located inside the relay with poles and the reed switch is located on the outside of the coil with a shaft that is parallel to that of the coil.

The realization of the design for a relay, manufactured in accordance with this design is not as complicated as the previous one, see Fig. 5.6. In the position in which it is illustrated in Fig. 5.6, the sensitivity of the relay is at its maximum. The desensitization of the relay is achieved by rotating the reed switch in such a way that an angle is formed between the longitudinal axes of the coil and the reed switch. The minimum sensitivity of the relay is achieved at an angle of 90° to the aforementioned axes.

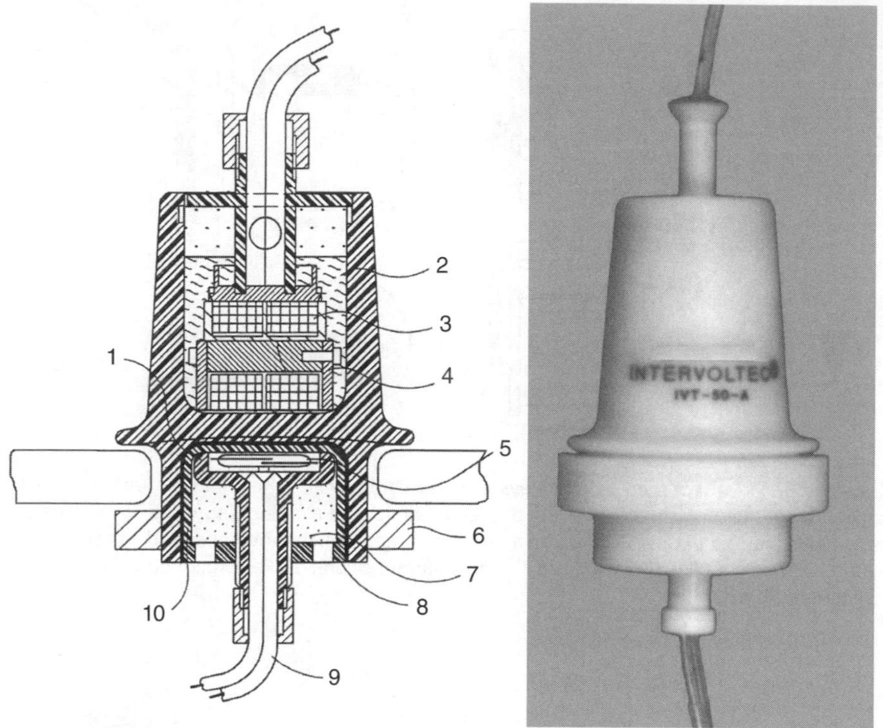

Fig. 5.6 Reed switch relay with a regulated pickup threshold, produced according to the structural layout b with a reed switch, located outside the coil, the longitudinal axis of which forms an angle with the longitudinal axis of the coil itself. 1 – A plastic, mushroom shaped capsule with a recess for the reed switch; 2 – A filled epoxy moulding compound; 3 – The coil; 4 – Ferromagnetic core; 5 – Reed switch; 6 – Nut; 7 – A filled epoxy moulding compound; 8 – A fixing nut in the rotating casing; 9 – The reed switch outputs; 10 – The rotating reed switch casing.

In this design regulation of the pickup threshold is achieved by rotating the casing 10 through 0–90° with the help of the tip of a capsule containing a reed switch that protrudes externally, and which is subsequently fixed in place by a fixing nut 8. The relay can either be attached to the external panel using the nut 6 (as per the illustration in Fig. 5.6) or with the help of a protruding flange with apertures and standard screws. This type of relay has a cylindrically shaped casing with a relatively large diameter (which exceeds the length of the reed switch tube) and the height, which is equivalent to approximately three times the length of the reed switch. The electrical strength of the insulation between the reed relay and the coil in this design of relay is significantly higher than that of the previous relay and can reach dozens of kilovolts. Suffice it to say that this design was realized by the author for a voltage of up to 70 kV (with a corresponding thickness of the insulation casing, its length, and the selection of corresponding insulation material for its manufacture).

The most compact relay with a regulated pickup threshold is that which is realized in accordance with the design illustrated in Fig. 5.7. In this relay the reed switch and the magnetic shunt are installed opposite each other eccentrically inside the rotating ampoule.

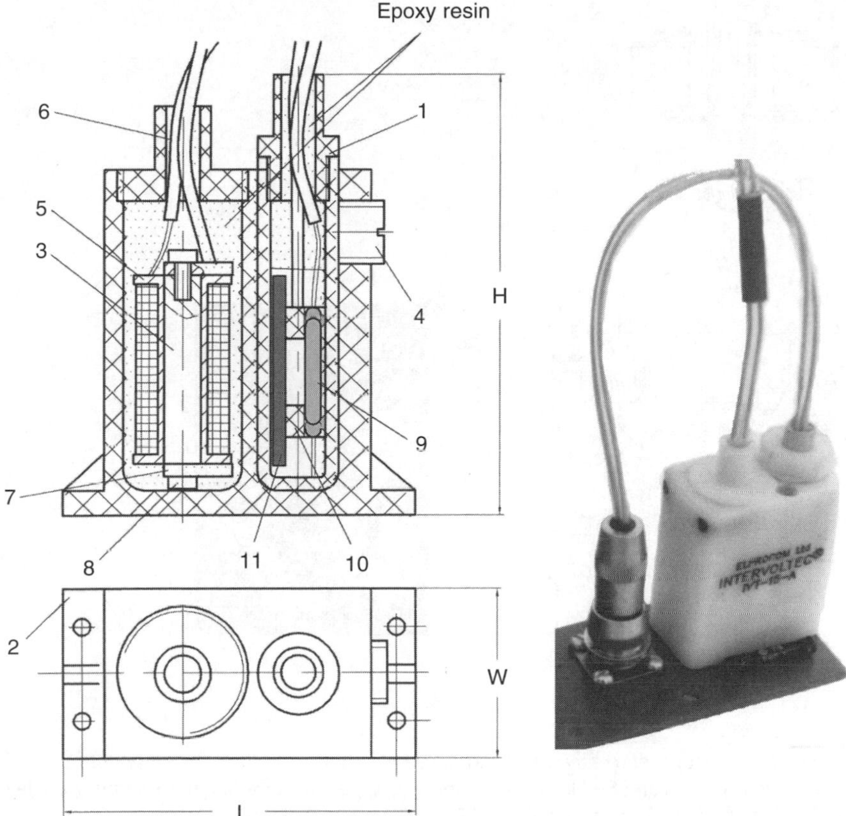

Fig. 5.7 The design for a compact relay with a regulated pickup threshold, that has been manufactured according to a design with an eccentric reed switch movement. 1 – The protruding section of the rotating ampoule casing. 2 – A mounting flange; 3 – Ferromagnetic core; 4 – Binding screw; 5 – A coil with a winding; 6 – Winding outputs; 7 – Core terminals; 8 – The mounting screws for the terminals; 9 – Reed switch; 10 – Insulation spacers; 11 – Magnetic shunt.

In the maximum sensitivity position, the reed switch should be the closest it can be to the core terminals in the control coil, and the magnetic shunt should be at the maximum distance. As the aforementioned ampoule is rotated the reed switch gets further away from the core terminals and its place is taken by the magnetic shunt, which weakens the magnetic flux around the reed switch. Use of this magnetic shunt enables a large degree of control of the activation pickup to be achieved given a small diameter for the rotating ampoule, that is to say it enables the dimensions of the relay to be reduced.

After tuning of the relay to the chosen pickup current the position of the ampoule is set by the screw 4. This relay also demonstrates an excellent electrical strength in the insulation between the reed relay and the control coil, especially when the reed switch and winding outputs are used in high-voltage insulation as a high-voltage cable

A differential relay with regulated pickup threshold can also be realized, which reacts to the variations in the values for the current and voltage, leading to the two different outputs for this relay, see Fig. 5.8.

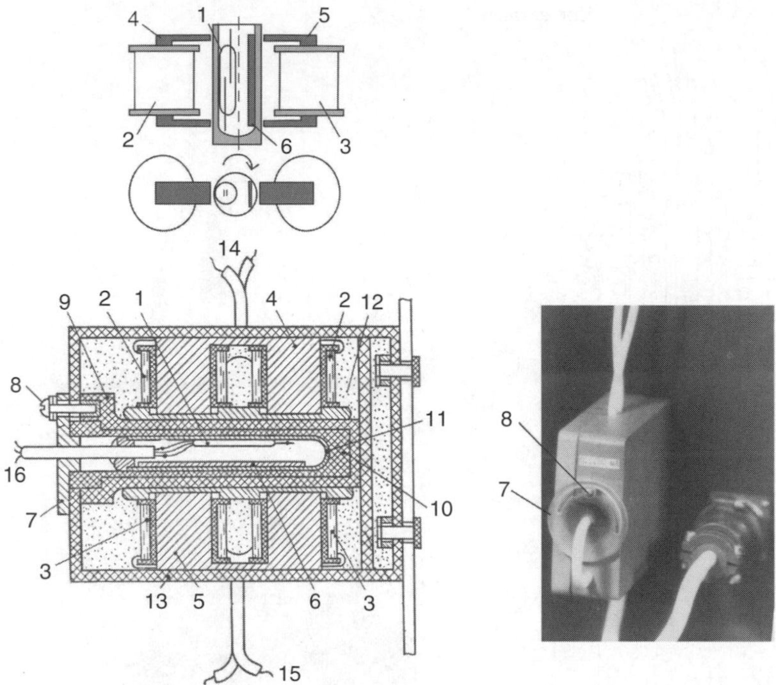

Fig. 5.8 A differential protective reed relay with a regulated pickup threshold. 1 – Reed switch; 2 and 3 – Coil with control windings; 4 and 5 – Flat, ferromagnetic U-type cores; 6 – Magnetic shunt; 7 –Relay tuning dial; 8 – Dial attachment; 9 – Static insulator; 10 – Rotating part of the insulator; 11 – The ampoule with the reed switch and a magnetic shunt; 12 – An epoxy moulding compound; 13 – The rectangular, plastic relay casing; 14 and 15 – The control winding outputs; 16 – The reed switch outputs.

The design for this relay in essence lies in its variety 'c' (Fig. 5.4) but it is defined by the presence of two coils, which are located in a single plane the other way around to the rotating ampoule with a reed switch and a magnetic shunt.

In this design as the dial rotates the mutual positioning of the reed switch 1 and the magnetic shunt 6 changes about the core terminals 4 and 5 in the control coils. In the process of rotation of the dial the relay gets further away from one of the coils and closer to a second, as a result of which the degree of influence of these coils on the reed switch (that is to say the output signals) changes. If the chosen polarity for the coils is reversed then if the rotating insulator 10 is in the central, neutral position the density of the magnetic field around the reed switch would be closer to zero. If the insulator were to rotate with the reed switch then the influence of the one coil on the relay would increase, while the other would weaken.

The design of this relay ensures a high degree of galvanic isolation between the inputs and outputs owing to the presence of a high-voltage insulator 9. If this insulator were moulded from a high-quality plastic together with the casing and if a high-quality epoxy compound were poured in following assembly of the relay in a vacuum, an electrical strength in the insulation could be attained with this design of dozens of kilovolts.

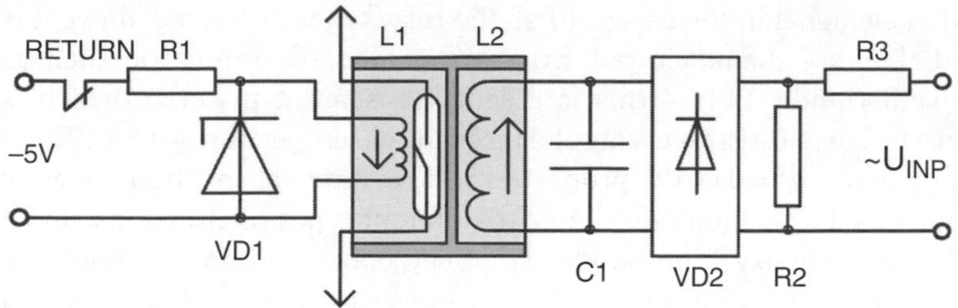

Fig. 5.9 A minimal voltage reed switch relay.

For a relay to work together with the DPR (as discussed at the start of the chapter) an insulation of dozens of kilovolts is, naturally, unnecessary. Insulation, however, of 5–10 kV of pulse voltage would not have any affect at all if this refers to a protection device that is designed to provide protection from the impact of a high-power electromagnetic pulse, which as we know is characterized by high-power induced voltages. In the b and c designs, realizing this level of insulation is not a problem since these designs were developed from the very beginning to operate at high voltages [5.9].

The designs described here have been tested in practice and demonstrated excellent characteristics both with maximum voltage and maximum current relays. However, in certain practical scenarios, for example, when they are used in DPR devices, which employ a distance protection function, a minimum voltage relay may be required.

In order to realize the functions in a minimal voltage relay an additional L1 winding must be placed directly on the reed switch with a relatively small number of turns, which according to the design would be connected to a stabilized 5 V voltage source, but the operational L2 winding, which is insulated from the reed switch, is connected by a VD2 diode rectifier and a C1 smoothing bulk capacitor, see Fig. 5.9, that possess great reserves of voltage (the output voltage from a standard voltage transformer used in electrical power systems does not normally exceed 100 V).

The L1 and L2 windings are connected in opposition, in such a way, that the total magnetic field around the reed switch is close to zero in standard operating mode. If the input voltage is drastically reduced (to which the distance protection of power lines would ordinarily react as the current increased) the magnetic field in the L2 winding would weaken, while the magnetic field generated by the L1 winding would remain unchanged. The resulting magnetic field around the reed switch increases and this activates the reed relay. The design of this same reed relay could be any of those examined previously.

The device can make use of the following type BY2000 (Diotec Semiconductor) diode, designed for a voltage of 2000 V and a current of 3 A (and a pulse current of 80 A) in a DO-201 type casing (with a diameter of 4.5 mm, and a length of 7.5 mm). The type MKP1T041007H00 (WIMA) capacitor, 1 μF, 1600 V, has the following dimensions: 24 × 45.5 × 41.5 mm. These elements are located on a printed board in the device outside of the relay casing. The use of these high-voltage elements given a comparatively low voltage, which enters the circuit design after it has passed through the current divider on the R2-R3 resistors (15–20 V) are necessary to ensure that the device has a high resilience to overvoltages, generated by high-power EMP.

In all the relays described here, the use of miniature vacuum reed switches is recommended, which are capable of withstanding a test voltage of not less than 1 kV and which have an intrinsic activation time of around 1 ms, see Table. 5.1.

With the insulation elements in the design of all the relay types described above it is recommended that they are manufactured from a moulded thermoplastic such as ULTEM-1000 (Polyetherimide, PEI) – this is a semi-transparent material, which is amber in colour and which possesses excellent total mechanical, temperature (−55 + 170 °C) and electrical (33 kV/mm, tgδ = 0.0012) properties, a low water absorption capacity (0.25% over the course of 24 h), a high resilience to different types of emissions, and a relatively good adhesion to epoxy compounds. The filled epoxy compound should be STYCAST 2651–40 (Emerson & Cumming) – this is a binary compound that is black in colour and which demonstrates excellent dielectric properties (18 kV/mm, tgδ = 0.02), a low water absorption capacity (0.1% in the course of 24 h), a wide operational temperature range (−75 + 175°C), a very high viscosity in a liquid form and a good adherence to metal and plastics. This compound demonstrates a linear expansion coefficient, which is close to that of ULTEM-1000, which is very important if a relay is to operate across a wide temperature range. CATALYST-11 should be used as a curing agent.

It is worth bearing in mind that a reed switch should never be soaked in an epoxy compound directly. It needs to be preliminarily coated in a damping material that compensates for the mechanical tension, which arise during the epoxy compound curing process.

In order to prevent the penetration of high-frequency and pulse interference into the output circuits of the relay via its capacitance, the reed switch has been placed inside an earthed, thin walled aluminium ampoule.

The technical solutions examined above could serve as a practical basis in the preparation for the manufacture of current and voltage relays with a regulated activation threshold, for devices that are designed to enhance the resilience of DPR to cyber-attacks and EMP.

5.3 Technical and Economic Aspects Affecting the Active Methods of Protecting DPR

The vulnerability of DPR to IDRT (both electromagnetic and cybernetic) was demonstrated previously, the need to protect DPR was established and the use of a specific method of protection was set out, based on the joint use of a DPR and a starting element based on a reed relay, functionally connected in series with a DPR and unblocking DPR only in cases in which just one of the controlled parameters approaches (current, voltage, angle between them, power etc.), the pickup threshold of the DPR, see Fig. 5.1.

The definition of the problem itself, as well as the proposed method of protecting DPR from IDRT are so unusual and so different from the knowledge gained so far that it invariably gives rise to a whole raft of questions among specialists as well as a flurry of emotions (alas not always positive). The lack of answers in articles that have been published previously often leads to confusion, and hence to the proposed method not being accepted at all. Therefore, we shall try to formulate the most frequently asked questions on this topic and provide answers to them.

Question 1. Judging by the design, are the reed switches hanging on the DPR on all sides, like garlands on a Christmas tree?

It is very obvious that the reed switches are not 'hanging like garlands' off the inputs and outputs of a DPR but along with all the other elements in the proposed protection device are housed inside a separate shielded casing, which is similar in design to DPR casings but with one exception – that there is no need for a screen in these cabinets although there is access to the pickup threshold control assemblies for the reed switch starting element. This separate module is fitted with the connector for connecting to external circuits as are fitted in DPR.

Question 2. There is a widespread opinion concerning the unreliability (specifically their sticking) of reed switches. To what extent is their use justified in a device that should demonstrate enhanced reliability?

The reed switches, or more specifically reed switch based relays that are used in the starting element differ from ordinary electromechanical relays in terms of a whole series of positive qualities. Firstly, the contact-elements in the dry reed switch are housed inside a sealed tube filled with a mixture of pressurized inert gases or in a vacuum and as such they are not subject to the influence of negative factors from the external environment (damp, dust or gases). These contacts do not require adjustment or cleaning throughout their entire service lives. Secondly these reed switches have a pickup time that is 3–5 times higher that of standard electromechanical relays. Thirdly, operating on an alternating current reed switches have a release ratio of 0.9–0.95, which surpasses similar parameters for standard relays. Fourthly, a level of galvanic isolation between the inputs and outputs (between the coil and the contacts) of dozens of kilovolts can easily be achieved in reed relays, which is not attainable using standard electromechanical relays. Fifthly, in contrast to standard relays, reed relays have a distinct and gradual pickup threshold when the current is gradually increased in the control coil, which enables sensitive protective measuring elements based on reed switches. In addition to this it can be said that dry reed switches are not sensitive to their position in a space and connect well with electronic, electromagnetic and magnetic elements, which enables a number of different functional modules and devices to be created based on these reed switches [5.10].

High quality vacuum and gas filled reed switches, which are manufactured by the leading companies that specialize in this field (and are the same switches that it is proposed would be used in the device [5.11]) are not cheap ($15–30 per piece), but are highly reliable components, that have been used widely not just in industry and in communication technology, but also in military and aeronautical technology. In view of many of their parameters reed switches occupy an intermediate position between semiconductor and electromechanical switching elements. Therefore, automatic telephone exchanges (ATEs) that are based on reed switches (such as the 'Quant' type stations and so on) are known as 'quasi-electronic'. According to the technical specifications the service life of these ATEs is set at 40 years, while the requirement for the failures in reed switches in that same time period should not exceed 0.3%. Just one of these requirements speaks for itself.

However, reed relays do have one principle difference from standard electromechanical relays: their magnetic system is not insulated from the contacts, but is formed by the contacts themselves. This difference is the reason behind the low overload capacity among reed relays. In contrast to standard relays, reed relays do not allow even a small current overload of the contacts to occur. The reason behind this lies in the fact that the magnetic field around the current that passes through the reed relays' closed contacts is directed counter to the magnetic field in the winding, which keeps the contacts in a closed position and weakens the magnetic field, thereby weakening the contact pressure, right up to forming a gap. This leads to increased erosion and sometimes to the contacts welding together, even if a current flows through them for a short time if it exceeds the maximum permitted value for these relays. Ignorance of these idiosyncrasies in reed relays and of how they are to be used without taking into account their differences from standard relays in terms of their overload capacity often leads to equipment failures, and hence to a lack of confidence in reed relays. If their operating mode is chosen correctly reed relays can ensure reliable circuit switching over millions of activation cycles. When reed relays are used to switch external circuits, in which the current can vary considerably, nobody wants to monitor the current mode in which the reed relays are operating. It is far easier not to use them at all, which often happens in practice. In the proposed design some of the reed relays are only wired into in the device's internal circuits, where the current overload is dozens of times less than the maximum permitted overload for reed relays. The rest of them de-energize discrete input circuits, in which the currents do not exceed a few milliamps, which is less than the maximum by a factor of two. It is only through reed switches, connected in series with the output contacts of a DPR and which are designed to switch the circuit breaker trip coil, that currents of up to several amps are able to flow. However, firstly, these reed switches do not conduct the switching of these currents directly, but only serve to switch the circuit without a current (so called 'cold switching'), and secondly the ones that are selected are Bestact R15Us produced by the Japanese company Yaskawa, which ensures excellent reserves of current.

Question 3. Contemporary DPRs combine 10–20 or more different functions in a single terminal. Does this mean that the proposed protection device should also contain this same number of input relays?

No, it does not mean that they should. The point being that all the different DPRs that are realized today in a single terminal are based on measurements of the current, voltage, and of the angle between them. Correspondingly the input relays in the proposed protection device should contain threshold elements for the current, voltage and the angle between them. The activation thresholds for all these elements should be less than the minimum level, selected as the threshold for the DPR.

Question 4. Why do expensive DPRs need to be used alongside some new and just as expensive protection devices, if it is possible just to resort to using cheap, electromechanical relays that are resilient to IDRT?

Yes, electromechanical protective relays have actually been used for more than a hundred years and up until now they have provided reliable protection from emergency modes in all kinds of electrical equipment.

Suffice it to say that such a large and widespread national energy system as the Russian system even today is comprised of almost 80% electromechanical protective relays. However, despite the fact that electromechanical relays have proved their excellent reliability, around 30–40 years ago all the leading global manufacturers stopped developing and upgrading electromechanical protective relays and began to develop first semiconductor relays and then DPRs intensively, which copied the functions of semiconductor relays, see Fig. 5.10. Only years later have DPRs appeared on the market with a wider selection of functions and which demonstrate enhanced characteristics.

Around 20–25 years ago the majority of leading global manufacturers of protective relays had just grown tired of producing electromechanical protective relays, having concentrated all their efforts on developing DPR. The main reason for this phenomenon was that producing printed boards with electronic elements on automatic equipment and then testing them also on automatic equipment was significantly cheaper than manufacturing a lot of miniature precise mechanical elements on highly accurate lathes and milling machines, and then manufacturing a sufficiently complicated mechanical construction from these elements by hand, and then testing and adjusting them, also by hand.

In view of the huge difference in the cost of production between an electromechanical relay and a DPR the consumer also wins since the cost of a DPR manufactured by the global leaders in relay manufacture is today much cheaper than an electromechanical protective relay with similar parameters. The assertion that an electromechanical protective relay is significantly cheaper today than a DPR is incorrect in the majority of cases, and does not bear up in an analysis of global market prices. Thus, for example, if a LZ31 three-stage electromechanical distance line protection relay (which is produced by ABB) would in today's prices cost around $30–35 000, then its microprocessor analogue with enhanced characteristics – such as a D30 (General Electric) relay today would cost just $7000, while a Chinese analogue such as the GTL-823 for example (which is produced by Guatong Electric) costs even less: $4000–5000.

As far as market prices in the former Soviet countries are concerned they have become very distorted and do not reflect the price ratio that exists today on the global market. For example, if the prices of time-over current relays, which are similar in terms of their design and their characteristics were to be compared: The Russian RT-80 and the American IAC (Fig. 5.10) then it turns out that the Russian manufactured relay (which costs around $60) is more than 20 times cheaper than the American IAC (which costs around $1400).

This difference in prices could be explained by the use in Russia of cheaper equipment, cheaper materials, and most importantly, cheaper labour. It is, however, to be expected that the difference between the cost of Russian and western manufactured DPR would be, if not exactly the same then at least close. What do we see though in practice? If the distance line protection relay: the aforementioned D30 relay (produced by General Electric) and the Sirius-3-LV-3 relay (produced by Russian enterprise Radius Automatica, see Fig. 5.11), which is similar in terms of its parameters were compared, it turns out that the costs of these two devices are more or less the same ($6500–7000). How can this be explained taking into account the previous information? Even if the fact that the Russian DPR makes use of a great deal of western manufactured electronic components is taken into account it would still be difficult to explain in objective terms such a strange price ratio. More than likely this represents a clear inflation of prices for Russian products with the aim of attaining bumper profits.

If the existence of a distorted price formation process is used as a basis, then there is every likelihood that this would give rise to serious difficulties in terms of using the proposed protection device in Russia.

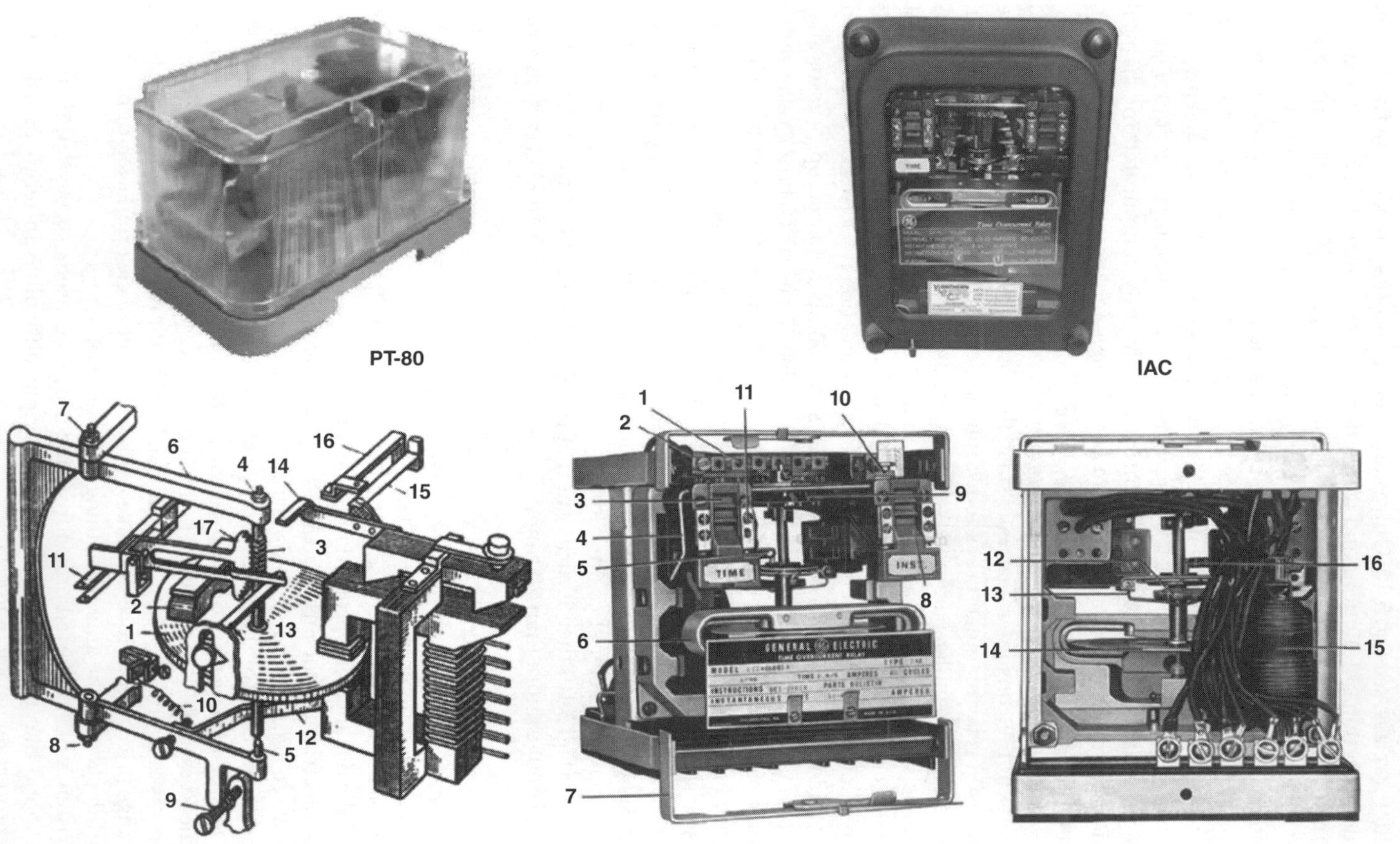

Fig. 5.10 The external view as well as equipment that is similar in terms of its parameters and design to electromechanical dependent time lag current relays: on the left is a RT-80 type produced by the Cheboksary Electrical Equipment Factory (Russia), while on the right is a IAC type produced by General Electric (USA).

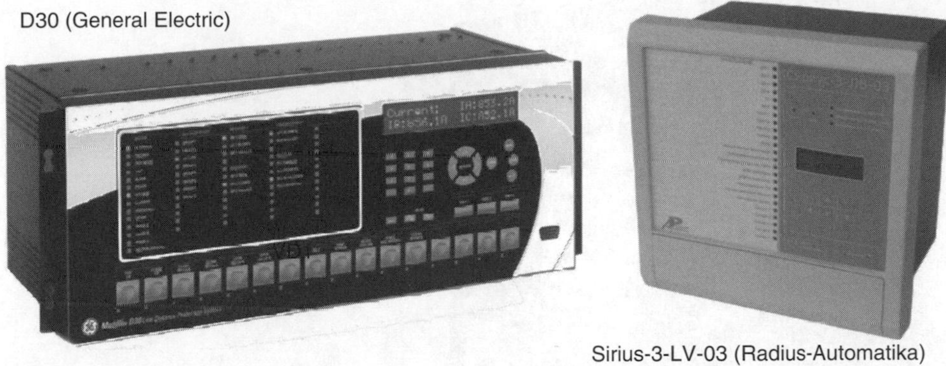

Fig. 5.11 The D-30 microprocessor based distance line protection relay (manufactured by GE, in the USA) and the Sirius-3-LV-03 (manufactured by the RPE 'Radius-Avtomatika' Russia), which are similar in terms of characteristics and cost.

On the other hand, the most powerful marketing campaign organized by the DPR manufacturers, developers, universities and research organizations that are interested in the financing of these projects has done its work. Today, to raise the question of a return to electromechanical protective relays is to be pariah in specialist circles and to be known as a retrograde that is attempting to halt technical progress. None of the specialists or the officials, who are responsible for taking the decision, would take on that responsibility. If they were to take it on, then it would be possible to assert that they would be subject to a furious stream of accusations of being a reactionary or of incompetence. Apart from that it has to be acknowledged out of objectivity that DPR do possess certain characteristics and functional capabilities that are unattainable for electromechanical protective relays.

Taking into account all these factors it can be stated that the question of returning to electromechanical protective relays is not on the agenda, even if it is economically justified given the price ratio prevalent in Russia.

Question 5. It is accepted that a return to electromechanical protective relays today is really out of the question. However, could the DPR be used in conjunction with these electromechanical protective relays instead of inventing new devices of some kind based on reed relays?

In actual fact the use of DPR and electromechanical protective relays in conjunction with each other does happen in practice (Fig. 5.12). It is true that these are not connected in series like in the proposed device, but in parallel, that is to say to duplicate one another with the aim of enhancing reliability. As was demonstrated previously [5.7] the method of using DPR and electromechanical protective relays together (that is to say to connect them in parallel) is by definition incorrect.

Using a parallel connection of this kind the electromechanical protective relay really should reflect all the functions of a DPR and have the same installations. In any case joint use of a multifunctional DPR and an electromechanical protective relay would require a whole collection of electromechanical protective relays, which are far from cheap, which makes this whole design very doubtful owing to its high cost and the large

Fig. 5.12 A fragment from a 160 kV panel for the distance protection of critical lines, which contains an LZ31 type electromechanical distance protective relay (top of the unit) and which is connected to operate in parallel with the MICOM P437 microprocessor based protection relay (bottom of the unit).

space required for the installation of a large number of different electromechanical protective relays.

Question 6. In order for it to be universal and work to its full potential the proposed protection device should in terms of its functional capabilities be the same as the collection of electromechanical protective relays. Does that mean that the cost of this device should be more or less the same? Why would it be cheaper?

Let us examine how an electromechanical protective relay works. Take for example an electromechanical time-over current relay, in which an aluminium disk begins to rotate as a certain current level is reached in the activation threshold, while the mobile contact, which is connected to this disk, comes closer to the immobile contact (an IAC type relay, Fig. 5.10). After a certain period of time, which is set by the speed of rotation of the disk, (which itself is defined by the level of the current flowing through the relay coil) the contact closes (via the auxiliary relay) the circuit breaker trip circuit. No time delay that is dependent on the current is required for the starting element in the proposed protective device. This starting element should only work at a certain level of current, which is somewhat less than the pickup current for the aforementioned disk. That is all. No additional functions of any kind are required of it, since all the other functions would be carried out by an activated DPR.

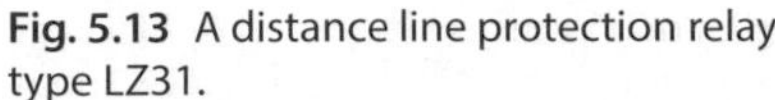

Fig. 5.13 A distance line protection relay type LZ31.

That is to say, that in this example instead of a complex and expensive time-over current relay a much simpler relay is used, which contains a coil and a reed switch only. As an additional example several types of distance line protection relays can now be examined.

The electromechanical variant of this relay, such as the LZ31 depicted in Figs 5.13 and 5.14 for example, contains a great number of complex, interlinked electromechanical assemblies, that provide three stages for measuring the resistance in the line including fault finding, that correspond to these three timed stages, and to the specific shape of the characteristics and so on. As mentioned previously, the cost of this relay is around $30–35 000.

Apart from that, this entire complex is started by the simplest of starting elements, which provides a balance control between the current and the voltage in the line, see Fig. 5.15. The activation of this starting element is achieved by disturbing the balance between the current and voltage.

In relatively complex and large distance protection relays, such as the RYZKB, RYZOE and RYZFB types, for example, that were manufactured by ASEA in the 1970s (Fig. 5.15), several protection functions are realized. However, all these relays have a relatively simple starting element, the design of which is illustrated in Fig. 5.16. These starting elements were an integral part of these complex designs and were not produced separately. An exception to this are certain types of relays, which were produced by Cheboksary Electrical Equipment Factory, such as the KRS-112 for example, illustrated in Fig. 5.16, which contain special inductors and four phase induction mechanisms with a rotating rotor. This relay is, in essence, a separate starting element for distance protection.

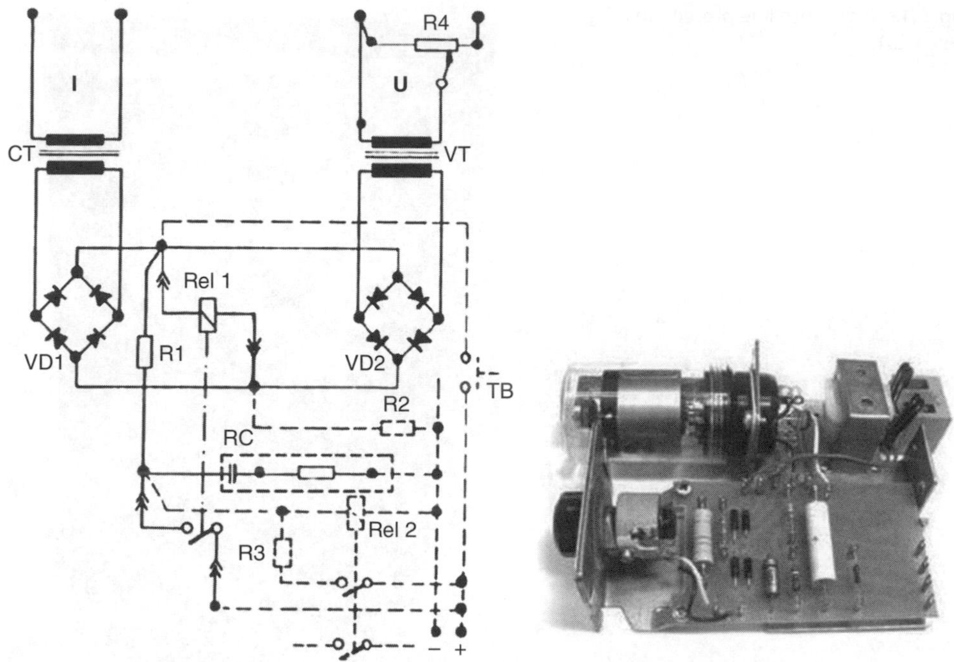

Fig. 5.14 The operating principle and the design of a starting element for an LZ31 distance protection relay.

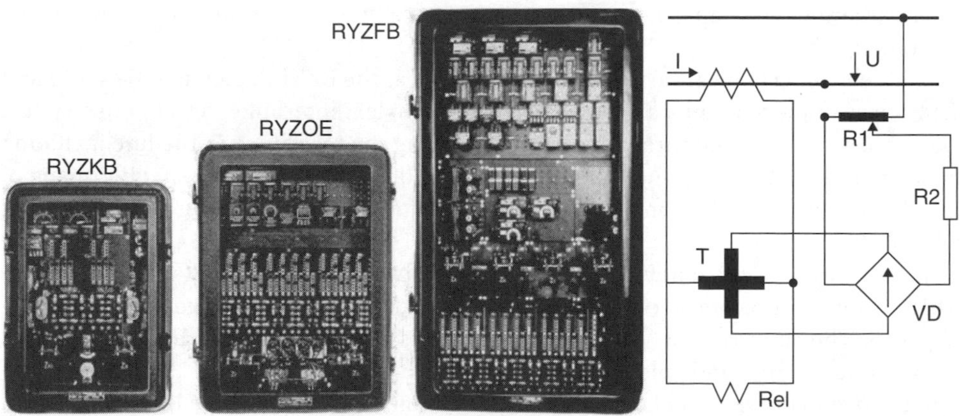

Fig. 5.15 An electromechanical distance protection relay such as those produced by ASEA as well as the circuit diagram of the SE (these were produced in the 1970s).

These relays, however, are very complex, expensive and have very large dimensions. The use of a design that is already morally obsolescent in conjunction with the most modern DPR though could hardly be described as a good idea.

In this respect the HZM (Westinghouse) type distance protection starting element would be much more attractive, see Fig. 5.17. This is a very simple device, containing a

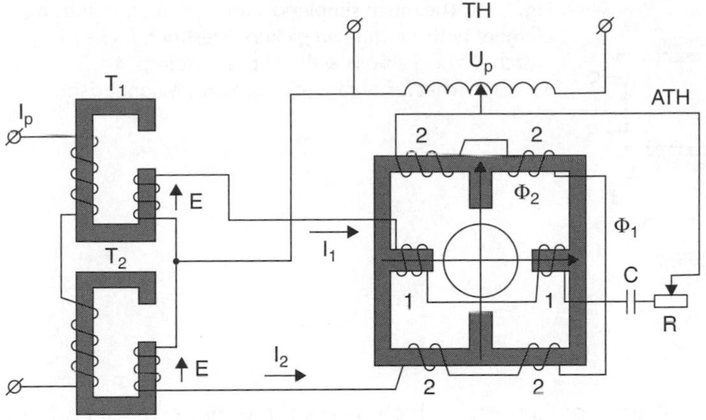

Fig. 5.16 A KRS-112 relay based on an induction mechanism.

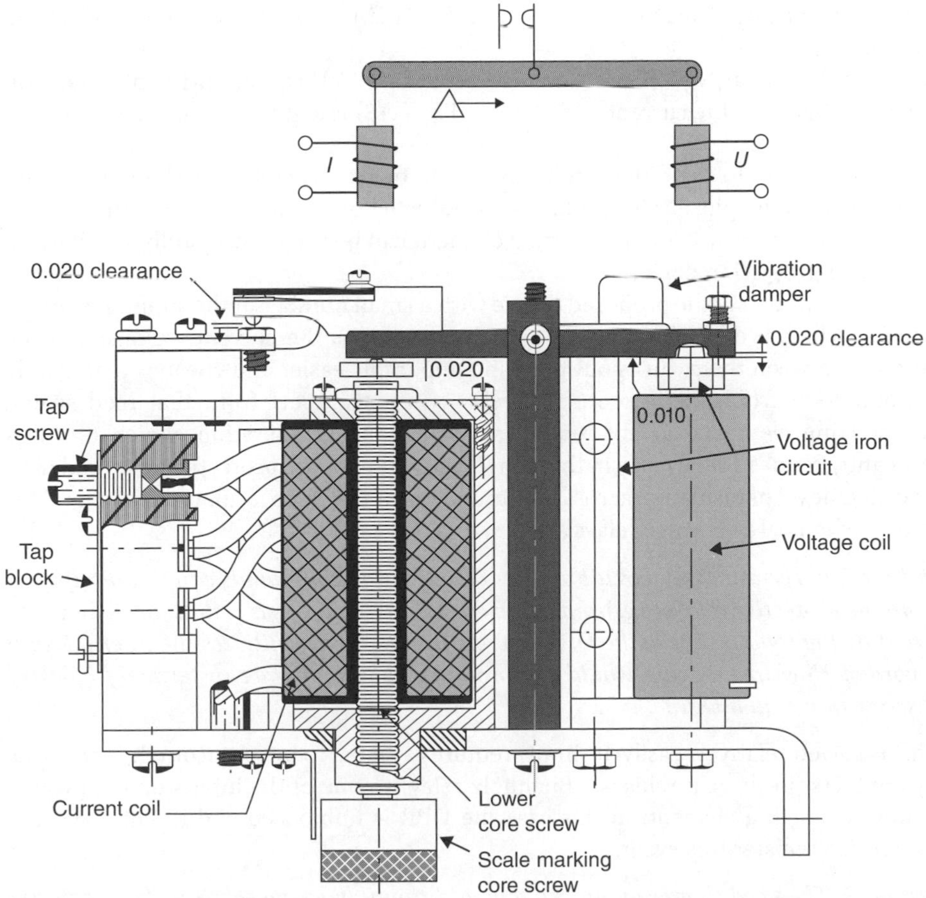

Fig. 5.17 A balance type electromagnetic starting element that is used in an HZM (Westinghouse) distance protection relay.

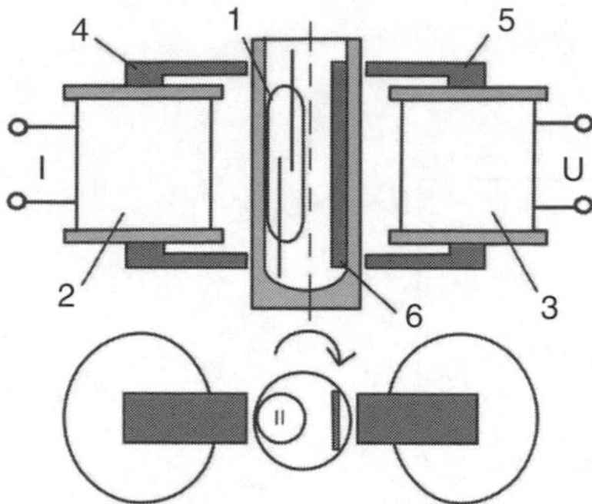

Fig. 5.18 The much simpler distance protection starting element with a regulated pickup threshold. 1 – Reed switch; 2 and 3 – Coils with control windings; 4 and 5 – П-form flat ferromagnetic cores; 6 – Magnetic shunt.

T-section core, and an oscillating rocker arm (which forms the upper cross section of the letter T) as well as two coils: a current and a voltage coil that act on the ends of the rocker arm. The position of this rocker arm, with the contacts attached to it is dependent on the balance of magnetic fields, generated by the current and voltage coils. This assembly is an integral part of the design of the HZM relay and was never produced separately.

A reed switch relay, which was designed according to this same principle based on the balance between the current and voltage (Fig. 5.18) is a great deal more simple and reliable [5.11].

This relay reacts to the difference in magnetic fields, generated by the current and voltage coils and its pickup threshold can be adjusted over a wide range as the capsule rotates with the reed switch. This starting element can be used successfully as a starting element in a protection device.

Using this approach, the proposed device with a small number of the simplest starting elements (which are based on a reed switch), as well as the current, voltage and the difference between them, turns out to be incomparably easier and cheaper than a fully functioning electromagnetic protective relay complex. Apart from that reed switch based starting elements do not require servicing during operation, which means a significantly reduced time delay in the overall activation of the protective relay and have an excellent level of insulation across inputs and outputs, which is unattainable for older electromechanical protective relays.

Question 7. In certain cases, circuit breaker deactivation commands can be sent directly from the protective relays (e.g. from a Buchholz relay) and this is simultaneously duplicated by the signals sent to the DPR discrete inputs, which initiates the internal fault recorder. How in this case would the proposed device operate in terms of blocking discrete inputs in a DPR?

This is solved relatively easily: it simply requires a signal to be sent from the activating relay contacts (in this case this is a Buchholz relay) to one of the inputs on the protection device starting element. In this case the DPR is unblocked and its internal fault recorder can register this event.

Question 8. The need to prevent any kind of additional blocking elements from entering the circuit breaker trip coil circuit is well-known, but in the proposed device this circuit is interrupted by a contact in an additional relay. Is this acceptable?

In fact, the normally closed contact in an additional relay is not connected to the circuit breaker trip coil circuit, but to a circuit that links the DPR output contact with the circuit breaker trip coil. The circuit breaker trip coil circuit remains free so that any other external contacts can be fitted to it, as well as hand operated keys.

Question 9. How does the system deal with complex protection systems, like those that contain a transformer magnetizing inrush current blocking scheme with filters for 2 and 5 harmonics, for example? Should the proposed device also contain a magnetizing inrush current blocking scheme and filters? Or to use another example: differential overhead line protection. How can the operation of the device be safeguarded in the event of an emergency mode within the protection zone only?

No, these filters are not required for the starting element to operate and no magnetizing inrush current blocking scheme is required. The starting element only de-blocked the DPR at magnetizing inrush current for a period of no longer than approximately 10 s. The algorithm in the DPR itself ensures that spurious pick-ups are blocked. Once the time period of 10 s has expired the starting element returns to its default and blocks the DPR once again. The same is also true of differential protection. As far as the starting element is concerned the location of the short circuit point within the protection zone or outside it – is not important. What is important is the presence of a short-circuit current that activates the starting element, and the DPR will define the damaged area after it has been de-blocked by the starting element. The activation period for the starting element is approximately 6 ms, which given a pick-up time for the DPR of 20–40 ms has almost no effect on the overall operating time for a protective relay.

Question 10. If the electromechanical protective relay and the DPR are connected in series the relay protection functions would be limited by the capabilities of the electromechanical relay, as an element with more moderate capabilities and inferior characteristics. Is this a good thing?

No, that is not the case. The proposed device does not in any way determine the properties or the characteristics of a protective relay. It simply activates the DPR as soon as just one of the parameters from the entire set of controlled parameters comes close to the settings on the DPR pick-ups. The subsequent behaviour of a protective relay as well as its reaction to an emergency mode will be set entirely by the properties and the characteristics of that DPR.

In practice, it is clear that more complicated operating modes can be found for DPRs, which have not been examined in this book and for which a specific starting element would need to be developed. However, even if a specific starting element were required then simpler, cheaper and faster acting elements can be created on the basis of a combination of reed switches and magnetic circuits than ordinary electromagnetic protective relays. For example, the device illustrated in Fig. 5.19 is perfectly acceptable to use to control the angle between the current and the voltage or as a power measuring element.

Additional possibilities arise when a combination of magnetic and high-voltage semiconductor elements equipped with a reed switch is used. For example, Fig. 5.19(a) illustrates the most basic design that reacts to the difference in currents, while Fig. 5.19(b). illustrates a device with a biasing characteristic (a coarsening of the sensitivity to the differential current in relation the direct current).

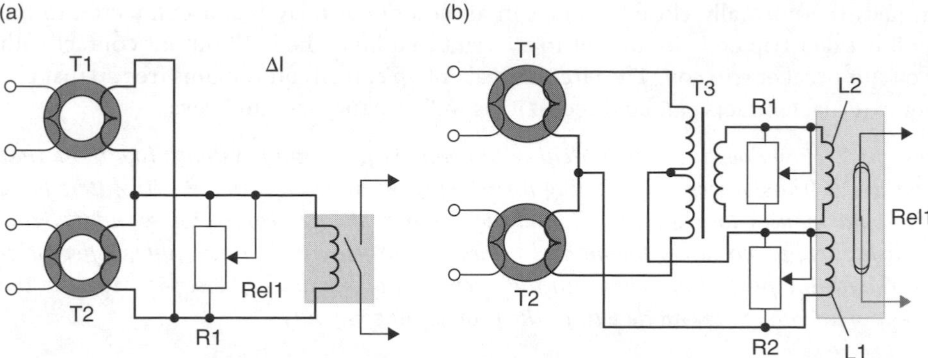

Fig. 5.19 Two variants of quasi-electronic differential protection starting elements. (a) The most basic design. (b) A device with a biasing characteristic.

Question 11. High-frequency and pulse interference can penetrate DPR both via the current and voltage circuits. How can they be protected using this proposed device?

The issue of the expediency of protecting the current and voltage circuits and of specific technical solutions to provide such protection requires additional study. The point being that the input currents and voltages enter the electronic circuitry of a DPR via the input current and voltage transformers that are found in each DPR, and which transform the sufficiently high-power input values into sufficiently weak signals. These signals represent units of volts that enter an analogue-digital converter (ADC) in which the analogue signals are quantized to a certain level and are transformed into digital code.

The quantization (digitalization) takes place inside the DPR with a sufficiently low frequency of 600–1200 Hz and as such the entire process takes a certain amount of time that is well in excess of the of the duration of the HPEM pulse. The ADC is simply not capable of carrying out the necessary transformations in the course of the duration of the HPEM. Therefore, the nature of the impact on the DPR of HPEM pulses, that penetrate through the current and voltage circuits would be defined by the inductive and conducting pick-ups that are not different to the disturbances that penetrate DPR in other ways. However, there is a way to significantly weaken these disturbances by disconnecting and shunting secondary current and voltage circuits in the aforementioned internal input transformers. To achieve this the circuits must be withdrawn to a separate connector by the DPR manufacturer so that the reed switches in this proposed protection device can be connected to these circuits.

Thus on the basis of the above analysis it is clear that the practical realization of the proposed method of protecting DPR from a technical and economic perspective is perfectly feasible, on the proviso that the functions of the DPR are kept separate from all other functions, which today hang off a DPR like garlands on a Christmas tree. Naturally enough, it is the DPR manufacturing enterprises that are able to offer consumers a quasi-electronic protection device as an additional option, in order to enhance security and reliability in the operation of assets that depend on relay protection that should bear the responsibility for this realization.

5.4 Protecting the Circuit Breaker Remote Control System

Owing to the negative trend in hanging all manner of additional functions that bear no relation to relay protection onto a DPR [5.12, 5.13] the realization of the measures proposed above to protect DPR would in certain cases become more complicated. This relates to the widespread use of DPR to provide a circuit breaker remote control system (RCS). It is very obvious that this use of a DPR has nothing to do with the functions of relay protection, and the remote connection of DPR along the communication channels with the aim of changing the position of circuit breakers is very hard to differentiate using hardware approaches from a cyber-attack.

As has been demonstrated many times previously, the task of enhancing the reliability of relay protection cannot be solved by combining the functions of a DPR with those that bear no relation to relay protection, such as the popular functions like monitoring the serviceability of electrical equipment and the remote control of circuit breakers. A DPR should be used exclusively for solving issues in relay protection. Moreover, there is an enormous number of specialized devices on the market today that are designed to solve other issues such as monitoring the serviceability of electrical equipment, from the most basic of relays, which control the integrity of the circuit breaker trip coil circuit, to the most complicated complexes, which measure the composition of gases formed in transformer oils, or the level of partial discharges in insulation in real time. In our opinion the RCS should also be separated from a protective relay and this function be conducted by other hardware. This is the only way to enhance the resilience of a protective relay and to ensure its effective protection from IDRT. This division of functions gives rise to an opportunity not only to provide highly effective protection of a DPR, but also to create a circuit breaker remote control system that is protected.

The proposed circuit breaker RCS in Fig. 5.20 is a hybrid and combines a microprocessor based controller with a network data transfer channel, as well as a cable channel with an electromechanical relay. The principle task for this system is to enhance its survivability and to maintain serviceability following HPEM. The general idea for this system is that any command to change the position of a circuit breaker, that is carried via the network channel should be confirmed by a short, remote activation of an electromechanical relay at a substation by sending a voltage to its coil via a standard control cable. This would therefore require use of an electromechanical relay and why should a fibre optic cable never be used as the confirmation channel?

The problem is that a fibre optic cable would not solve the issue of protection from HPEM, since these fibre optic cables are fitted on both sides with complex, microprocessor based multiplexers that ensure the conversion of electrical signals to light signals at one end of the fibre optic cable and their restoration to electrical signals from optic signals at the other end. As our research into several different types of multiplexers has shown [5.14] they are not able to withstand even standard pulse overvoltages in accordance with the requirement for electromagnetic compatibility (EMC). If the internal electronic components in the multiplexers are damaged as a result of HPEM the condition of their output circuits becomes unpredictable.

Apart from that, since laying dedicated fibre optic cables and installing the interface equipment is relatively expensive there is a trend today to reject the use of dedicated fibre optic cables and to use existing computer networks based on cheap, twisted pair

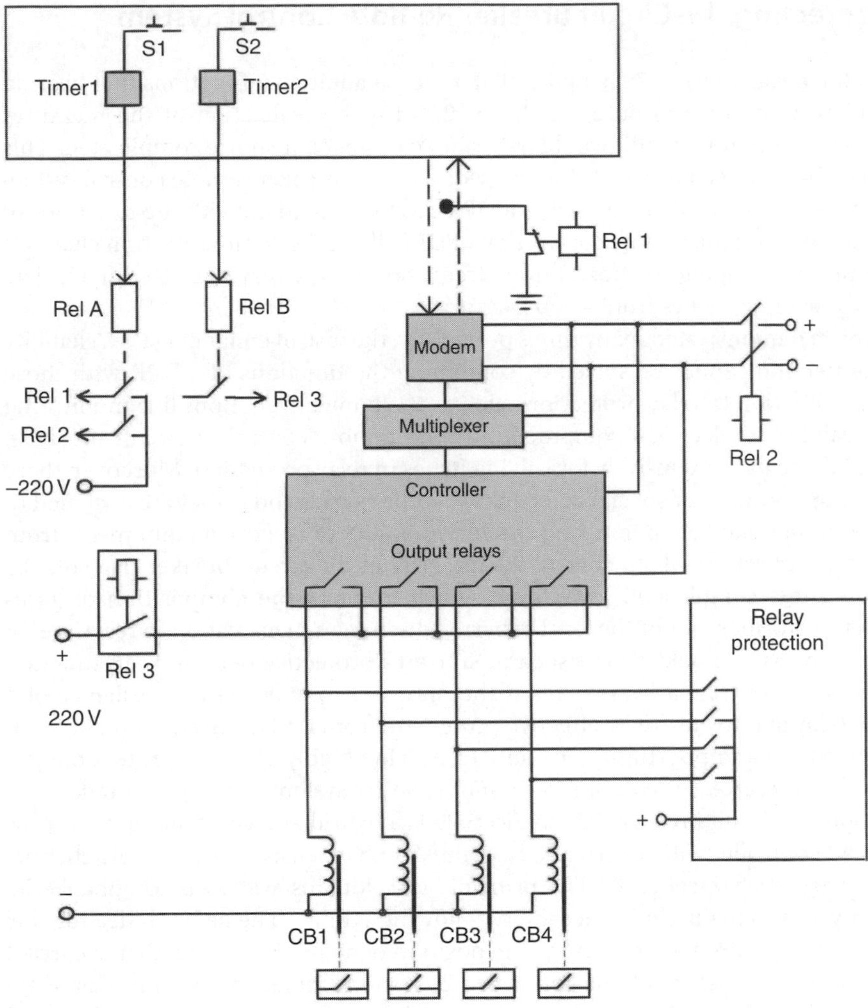

Fig. 5.20 The proposed structure for a protected circuit breaker remote control system. The feed circuits are shown schematically to simplify the diagram.

based cables. Furthermore, with the aim of making the control system, relay protection and automation even cheaper, the transfer to wireless Wi-Fi technology is undergoing serious examination. In any case many global leaders in the production of DPR are already producing them with built in Wi-Fi modems.

Technically speaking the idea of transferring the entire electrical power engineering equipment to communicating via standard computer networks, including wireless networks is the central idea behind the concept of 'Smart Grid'. In connection with this the need to develop special hardware based protection systems, and relay protection systems, as well as to protect remote control switching apparatus from IDRT, including electromagnetic attacks and cyber-attacks, that are not linked to computer networks and which demonstrate enhanced resilience to HPEM grows more acute.

This is the reason why we have chosen electromechanical relays, which control the operational currents in the control cables. Conductors are used, which belong to different control cables, with the aim of protecting this auxiliary communication channel from malicious external connections and unsanctioned activation of the electromechanical relays, and instead of a single relay two RelA and RelB are used, as per Fig. 5.20. Naturally the coils for these relays and conductors for the feed cables should be protected (by using TVS-diodes for example) from pulse overvoltages, which can be inducted in these conductors under the influence of a high-power electromagnetic HPEM pulse. Apart from that it is desirable to feed these relays using alternating current at an industrial frequency with a capacitor, inserted into the feed circuit and to use an isolation transformer alongside the substation to prevent relay pickups from a quasi-DC component (E3 component) of HEMP.

In such cases, when use of a control cable to control Re1A and Re1B relays is not possible in any circumstances owing to the significant distance between the dispatch point and the substation, a fibre optic cable can be used as an enabling communication channel. Furthermore, it is worth bearing in mind the reduction in the resilience of the systems to HPEM. In order to prevent the inadvertent sending of a signal to change the position of circuit breakers as a result of damage to the system's electronic devices they should be fitted with a self-diagnostics system. The fibre optic cable channel should be fitted with a permanent serviceability monitor to measure its own serviceability as well as the position of the RelA and RelB relays, and the controller should be fitted with an internal self-diagnostics system, which is activated automatically on activation of the Rel1 and Rel2 relays and which should also incorporate an interrogator to assess the condition of the output relays (these should be included) and the serviceability of the communication channel. If a failure is discovered the self-diagnostic system should block any further operation of the controller. Apart from that in order to protect the system from HPEM different passive protection methods, discussed in Chapter 4, should be employed. Only when the dispatch point receives a signal concerning the good condition of all the elements of the system, can use of the system as a circuit breaker remote control system be authorized.

In the proposed device any command to change the position of a circuit breaker, which is sent via any kind of network channel should be accompanied by the short, remote activation of the two Rel1A and Rel1B relays along the control (or perhaps the optic) cable. The contacts in these relays activate the local, electromechanical relays: Rel1 (which unblocks the network communication channel), Rel2 (which feeds the electronic devices in the system), and Rel3 (which activates the circuit breaker feed coils). All these local relays can differ in terms of their characteristics. For example, the Rel1 is a high-frequency relay, while Rel3 is a relay with high-power contacts, which are designed for switching the direct current inductive loading. The presence of the two control relays RelA and RelB with separate communication control channels enhances the resilience of the system to unsanctioned access.

The RelA relay is the first to activate, and after the necessary information has been passed to the controller concerning the position of one or another of the circuit breakers and the closing of the contacts of the corresponding output relay in the controller, the RelB contact actives and its contact activates Rel3. The amount of time at which the RelA relay is activated is automatically limited by a timer, in order to prevent these two relays being activated permanently in error by staff. In reality this is a short period of

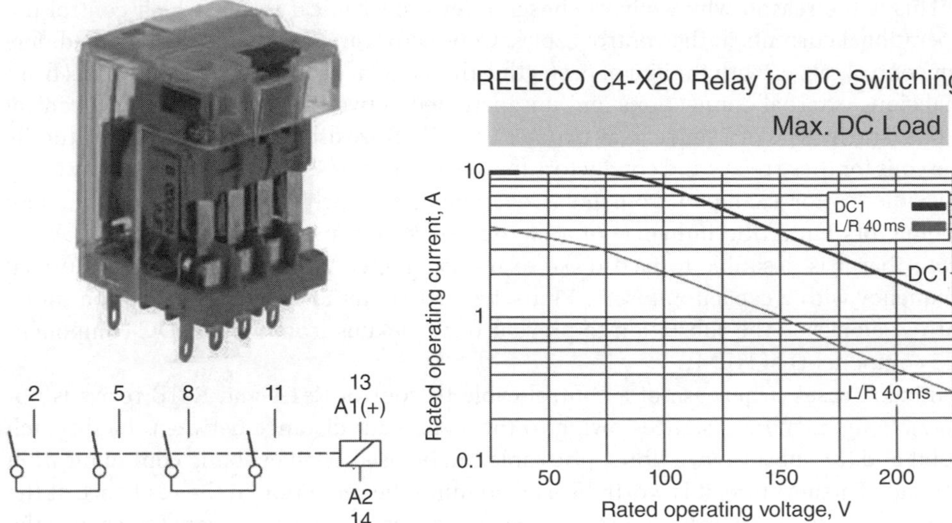

Fig. 5.21 A C4-X20 type relay produced by RELECO (with its cover partially removed), which has two contacts and a double break together with its direct current switching capabilities.

time, during which carrying out an effective cyber-attack is practically impossible. Blocking of the communication channel and closing off the feed to the controller outside of this short period of time excludes the possibility of preliminary activation of the controller's output relay as a result of a cyber-attack, with a subsequent spurious change in the position of the circuit breakers at the moment that the electromagnetic RelB relay is activated. These same measures also dramatically reduce the likelihood of damage to the sensitive electronic apparatus (the modem, multiplexer or controller) caused during HPEM.

Following a registered cyber-attack or an electromagnetic pulse remote control of the circuit breakers should be forbidden pending a special inspection, since the condition of the controller following these impacts is not known. The controller's output relays can be low-powered, standard relays, with which standard controllers are equipped. The Rel3 relay should have contacts that are capable of switching a relatively high-powered direct current load that is inductive in nature (such as the circuit breaker control winding) at a voltage of 220 VDC.

An analysis of the specifications for the widespread types of electromagnetic relays reveals that the majority of them are not designed for switching (even for turn ON) of direct current inductive loads at a voltage 220 VDC [5.15]. This purpose is served by specially designed relays: these provide multiple subsequent gaps in the switching circuit (Fig. 5.21) or contain a permanent magnet close to the contacts, which is designed to strip the electrical arc from the gap between the contacts (Fig. 5.22).

There are also relays with triple gaps per contact, see Fig. 5.23, which enables the trip coil in the old style high-voltage circuit breakers with a high consumption current to be controlled.

As can be seen in both cases, that is to say for the protection of a DPR and the protection of the circuit breaker remote control system, electromechanical relays are used.

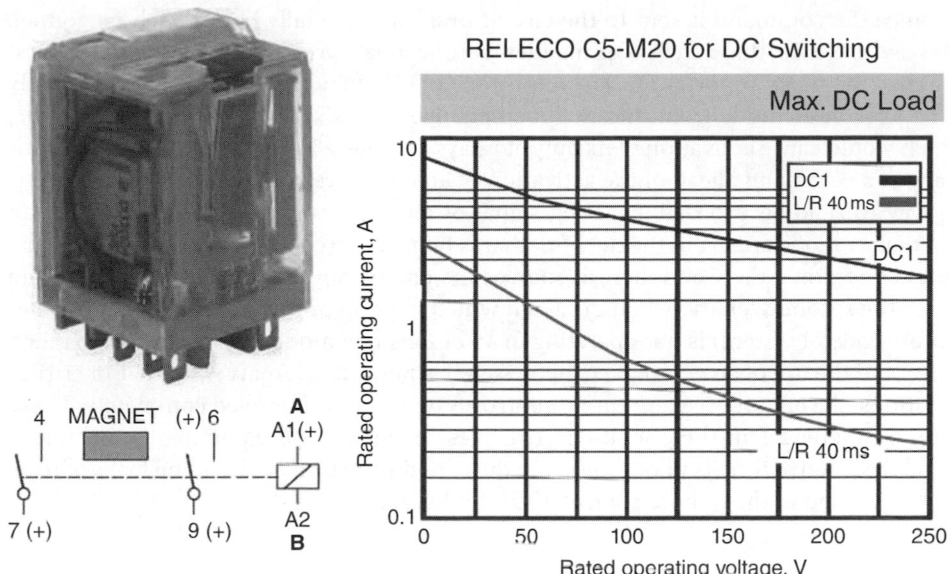

Fig. 5.22 A C5-M20 type relay produced by RELECO with two closing contacts and an arc-suppressing magnet together with the switching capability for the inductive load.

Fig. 5.23 An RMEA-FT-1 relay with a single closing contact with a triple gap, which is capable of switching a direct current of up to 3A with an inductive load at a voltage of 220V (the manufacturer is: RELEQUICK S.A.).

However, the use of these relays is different, which is linked to the different algorithms that DPRs and circuit breaker remote control systems use. If in the first instance the command is sent to the circuit breaker automatically when the mode of the electrical circuit or electrical equipment that is being controlled changes, then in the second instance the command is sent to the circuit breakers manually by dispatch personnel. This is linked to the different principles behind relay protection. Thus, in the first instance the most important task is to protect a DPR, which is operating permanently in an automatic mode from unsanctioned changes to its settings or its internal logic, which would cause activation of its output relays and there is no way to check the legitimacy of a command ahead of the activation of an output relay. Apart from that there is no way to send an external, enabling signal of any kind to a DPR in the event of an emergency mode arising in the network that is being controlled and this enabling signal should be formed there and then, as soon as the emergency mode comes to light. Then as per the second scenario when the asset, which is being protected (the circuit breaker remote control system) is not operating in an emergency mode, this task is made much easier and the use of an external enabling signal is permitted. Apart from that in critical situations, the circuit breaker remote control system can be annulled completely. These natural differences in the operating principles for protecting equipment from intentional destructive threats underline again the expediency of separating the tasks of relay protection and of the remote control of circuit breakers.

References

5.1 Gurevich V.I. Issues of philosophy in relay protection - *The World of Technology and Engineering*, 2013, No. 1, pp. 56–58.

5.2 Gurevich V.I. The cyber weapon versus power engineering - *PRO Electricity*, 2011, pp. 26–29.

5.3 Gurevich V.I. Problems of the electromagnetic impact on DPR, Ch. 1 - *Components and Technology*, 2010, pp. 60–64.

5.4 The impact of implementing cyber security requirements using IEC 61850 - *CIGRE Working Group*, the B5.38, August 2010.

5.5 Gurevich V.I. Intellectual networks: New perspectives or new problems? - *The Electrical Technology Market*, 2010, No. 6 (Ch. 1); 2011, No. 1 (Ch. 2).

5.6 Gurevich V.I. On certain approaches to finding a solution to the problem of the electromagnetic compatibility of protective relays in electrical power engineering. - *Industrial Power Engineering*, 1996, No. 3, pp. 25–27.

5.7 Gurevich V.I. Electromechanical and DPR. Is symbiosis possible? - Relay Protection and Automation, 2013, No. 2, pp. 75–77.

5.8 Gurevich V.I. A protective relay protection device - *Control Engineering Russia*, 2013, No. 3, pp. 47–51.

5.9 Gurevich V.I. *Protection Devices and Systems for High-Voltage Applications.* - Marcel Dekker, New York, 2003, 292 p.

5.10 Gurevich V.I. *Electronic Devices on Discrete Components for Industrial and Power Engineering* - CRC Press (Taylor & Francis Group) Boca Raton - London - New York, 2008, 419 p.

5.11 Gurevich V.I. Sealed contact reed relays with a regulated activation threshold - *Components and Technology*, 2013, No. 11, pp. 30–33.

5.12 Gurevich V.I. The technical process in relay protection: Dangerous trends in the development of relay protection and automation - *The Electro-Technical News*, 2011, No. 5, pp. 38–40.

5.13 Gurevich V.I. On multifunctional relay protection - *PRO Electricity*, 2012, No. 42–43, pp. 45–48.
5.14 Gurevich V.I. Current problems in relay protection: An alternative view - *The Electrical Power Engineering News*, 2010, No. 3, pp. 30–43.
5.15 Gurevich V.I. The idiosyncrasies of relays that are designed to de-energise high-voltage circuit breaker trip coils - *Electricity*, 2008, No. 11, pp. 22–29.

6

Testing the DPR Immunity to HPEM

6.1 An Analysis of Sources of HPEM

The requirements for the immunity of protective relays (including DPR) to electromagnetic threats is set out in the International Electrotechnical Commission (IEC) standards, series 60255. The general requirements for electromagnetic compatibility (EMC) for electronic apparatus are set out in the IEC standards, series 61000. Standards, which are identical to the IEC 61000 are used in Russia. However, all these standards relate to so-called "unintentional electromagnetic threats," that is to say to naturally occurring impacts (interference). The intentional electromagnetic threats, and specifically HPEM have a significantly greater impact on apparatus that was originally envisaged in the standard EMC requirements and as such the standard EMC requirements cannot be used in this case.

In order to develop specific recommendations and technical requirements for testing DPR for their immunity to HPEM certain tasks need to be resolved:

1) Classifying the types of HPEM and summarizing their technical parameters
2) Evaluating the parameters of HPEM acting upon a DPR in actual operating conditions
3) On the basis of the analysis of existing standards in the HPEM field, formulating technical requirements for the necessary equipment to simulate HPEM and test the resilience of a DPR to the impact of HPEM.
4) Carry out an analysis of the market for equipment designed to test DPR immunity.

As can be seen in Fig. 6.1. lightning has a far larger spectral density of electromagnetic emissions than even a high-power source of emissions such as an electromagnetic pulse from a nuclear explosion. However it is worth bearing in mind that all the energy in lightning is concentrated in a so-called stepped leader stroke and has a very focused impact zone, while HEMP covers a wider surface area, Fig. 6.2.

Each of these classes of HPEM can be divided in turn into individual impact types. HEMP encompasses three different impact types, in accordance with the three components of an electromagnetic pulse: E1 (early-time), E2 (intermediate-time), and E3 (late-time), described in Chapter 2.

Since this chapter covers testing of electronic apparatus, our interest will subsequently be focused on just the E1 component, as the most powerful and the most dangerous for electronic equipment.

Protection of Substation Critical Equipment Against Intentional Electromagnetic Threats,
First Edition. Vladimir Gurevich.

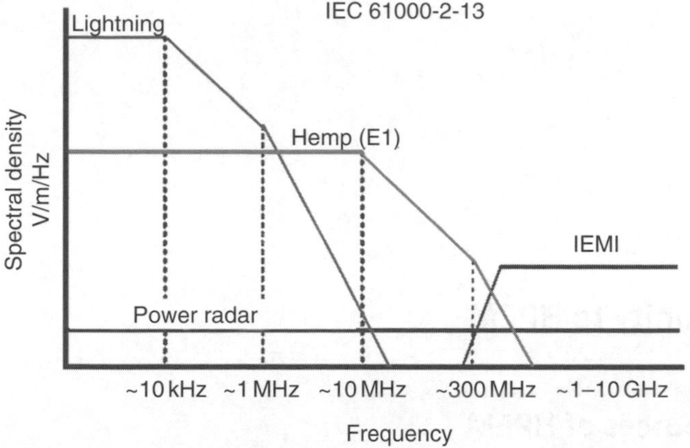

Fig. 6.1 The spectral density of emissions from different sources of HPEM in accordance with the IEC 61000-2-13 standard.

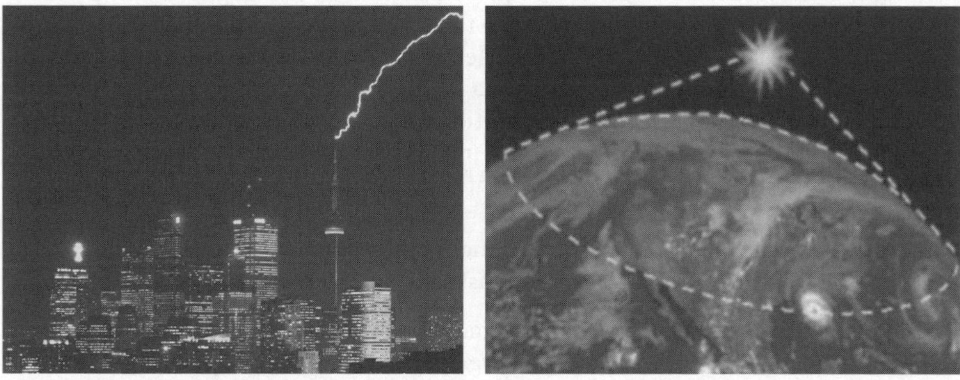

Fig. 6.2 The impact zones of lightning and HEMP.

Intentional electromagnetic interference - IEMI (see Chapter 2) can be divided in turn into two different types:

1) Unidirectional narrow-band or High Power Microwave emissions. These emissions can be emitted from:

   Generators that are operating continuously at a fixed frequency;

   Generators emitting packets of high-frequency pulses, with frequencies ranging from hundreds of hertz to dozens of kilohertz (Fig. 6.3.);

   Generators emitting an ultra-wide frequency range from dozens of megahertz to hundreds of GHz;

   Generators that are emitting high-frequency signals with a fading amplitude (Fig. 6.3.).

   Pulse generators with a peak emitted power from dozens of megawatts to units of gigawatts. These generators generate very short pulses in the nanosecond (from units to tens of nanoseconds) and sub-nanosecond range (with a pulse duration of tens of hundreds of picoseconds, with a frequency pulse repetition 0.1 to 10 KHz or more.

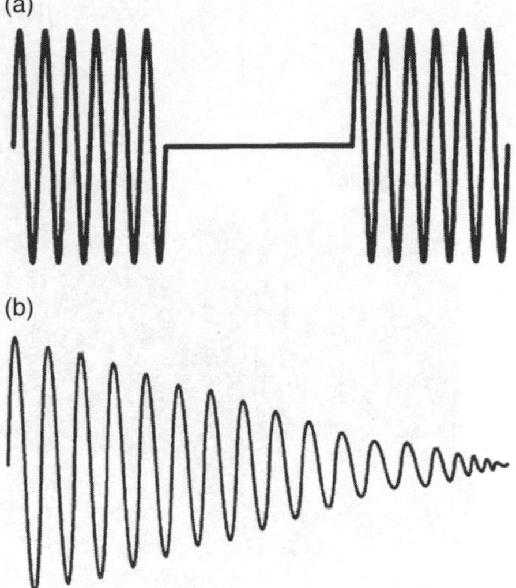

Fig. 6.3 Some of the different types of IEMI source signals, (a) Packets of high-frequency pulses with equal amplitude; (b) High-frequency pulses with a fading amplitude, repeating with a frequency of several kilohertz.

2) Electromagnetic pulse emissions from explosive sources of radio-frequency electromagnetic emissions (RFE), the majority of which are based on the use of an explosively pumped flux compression generator, generating a high power (1 GW or more) short pulse with a duration of less than 1 ns with a frequency of the resulting emissions ranging from hundreds of megahertz to hundreds of gigahertz in a single pulse.

In contrast to the classifications set out above the NATO AECTP 250 Ed 2.0: 2011 Leaflet 257 - High Power Microwave standard divides IEMI into four different types according to a different principle:

a) A mobile source of RFE, directed towards the target under attack and at a target that is outside of a secured area, but which is able to come close enough to the target to be able to hit it effectively.
b) A portable source of RFE, which can be delivered directly into a protected area underneath a person's clothes or in a briefcase and which can be placed right next to the asset under attack. A low-power source could be much more dangerous than a remote, high-power source of RFE, since its emissions are not weakened by anything.
c) A contact source of RFE, which injects energy directly into electrical wiring and cables, which are linked to electronic apparatus that is under attack, such as in a communication cable for instance. This source can be located both within a protected zone, or outside of it.
d) An explosive source of RFE that is manufactured as an electromagnetic bomb or missile, which emits a short, electromagnetic pulse, which can penetrate apparatus under attack through walls, and windows, as well as through wiring and cables that extend beyond the boundaries of a building.

For the sources of RFE listed above the NATO standard gives a maximum emitted power for the antenna of 1GW/m^2 and this is restricted by the electrical strength of

Fig. 6.4 The ultra wide band pulse generator with a maximum emitted power of 3.4 GW, which was developed at the Institute of High-Current Electronics of the Siberian Branch of the Russian Academy of Sciences.

air, which is equal to 1 MV/m. It is worth noting that in many other literary sources another value is given for the electrical breakdown voltage of air in normal climatic conditions: 2-3 MV/m. Apart from that even in non uniform electromagnetic fields the pulse discharge voltage of air increases further and surpasses the standard value for an alternating voltage of 50 Hz. The relationship between the amplitude of a pulse breakdown voltage and a voltage with a frequency of 50 Hz is known as the pulse coefficient k_i. For non-uniform electrical fields $ki = 1.1$-1.3. It is however worth taking into consideration that in a reduced air pressure (in high mountainous regions) or if there is dust or drops of moisture (mists) in the air the electrical insulation of the air can be reduced significantly. A reduction in the electrical insulation of air is also observed in the process of the ionisation of air under the influence of a high-frequency discharge. Nevertheless in 2008 at the Institute of High-Current Electronics of the Siberian Branch of the Russian Academy of Sciences a very wide band pulse generator was built recently, Fig. 6.4. with a maximum peak power of 3.4 GW, which significantly surpasses the limit on the emitted power indicated in the NATO standard. This generator operates at voltages ranging from -205 kV to + 157 kV and generates pulses with a duration of around 1 ns and a pulse repetition rate of 100 GHz. Meanwhile, in a laboratory in the United States located at Kirtland Air Force Base (in Albuquerque, New Mexico) a source of RFE was developed with an emitted power of 7.5 GW which was based on a so-called "vircator" (a generator with a virtual cathode) and which operates at a voltage of 4 MV and a current of 80 kA [6.1]. Since that time the design of the vircator has been significantly updated, and the power has been increased up to

40 GW, and the efficiency factor has been improved. This in turn has meant that a high-power RFE generator based on a vircator can be installed in the warhead of a cruise missile, which in the course of its flight is capable of burning out all the terrestrial microelectronics and computer based equipment that does not have any special protection.

This secret Boeing project (CHAMP), which began in 2008 [6.2] was a great success and was broadcast by the media, which characterised this success as the onset of a new era in future wars.

6.2 The Parameters of Testing DPR for Immunity to HEMP

In accordance with the 61000-4-25 standard tests for the immunity of electronic apparatus to HEMP should contain two component parts: tests for immunity to electromagnetic emissions (EE) and to the impacts of conducted interferences (CI). In turn CI are divided into two different types: pulse voltages, applied to the input points of an apparatus, and pulse currents inducted in long distance wiring and cables.

The definition of the specific testing standards begins with selecting one of 6 tests concepts. The 610000-2-11 and 61000-5-3 standards define these concepts. For DPR located in principle reinforced concrete or brick buildings fitted with protection from lightning strikes and without special protection filters the concept number 2b can be selected. This concept envisages the attenuation of the level of electromagnetic emissions down to 20 dB within a frequency range of 100 KHz - 30 MHz. For the chosen concept and the E1 component the electric field strength acting on the object being tested is set at 5 kV/m (level R4), while the intensity of the magnetic field is 13.3 A/m.

By way of a comparison: for wooden buildings, for which the electromagnetic emissions are not attenuated the electrical field strength would be 50 kV/m (level R7).

For this concept and for the E2 component the electrical field strength is set at 10 V/m and the magnetic field at 0.08 A/m. The parameters of the pulse EE described in the 61000-2-9, 61000-2-10, and MIL-STD-461 F standards are: the rise up time for the pulse (for the leading edge) is 2.5 ns, the width of the pulse is 25 ns, the shape of the pulse corresponds to that seen in Fig. 6.5.

In table 1 for this standard the amplitude of the test voltage for HEMP (which is marked as "special") is marked with an "X" and corresponds to levels E8 or E9.

In the next step a test impact level is chosen for the impacts of CI in accordance with the 61000-4-25 standard. For the chosen number b2 concept and the presence of wiring that is connected to the object under examination that has not been buried in the ground a test impact level of E8 is selected (this is to ensure a normal 50% probability of the immunity of the object) or E9 (for a 99% probability). The level E8 assumes an immunity for the object being tested to pulse voltages of 8 kV, while level E9 - 16 kV. The probability of 50% is considered normal for this standard and can be used for civilian equipment.

An Electrical Fast Transient (EFT) is assumed to be the test pulse voltage for the impacts of CI - this is a fast pulse, the parameters of which (apart from the test voltage amplitude) as well as the method of testing are set out in the IEC 61000-4-4 standard, Fig. 6.6.

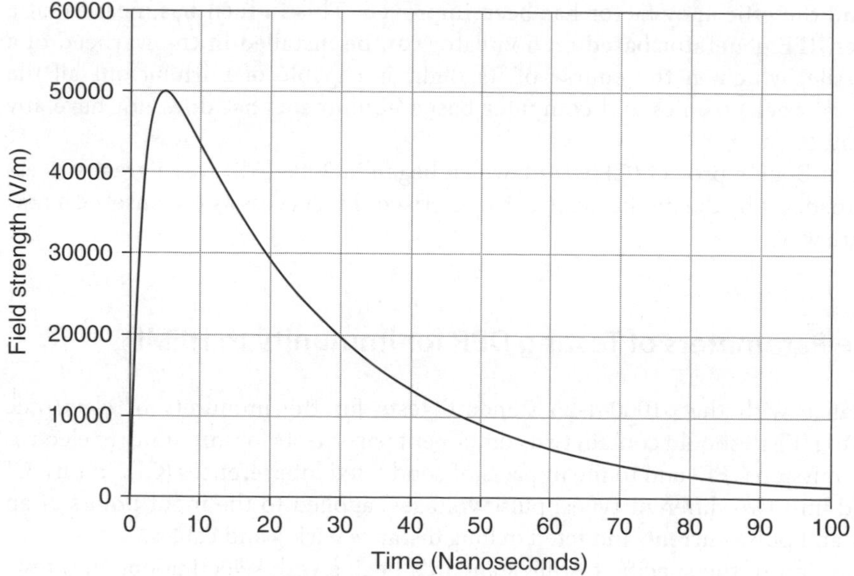

Fig. 6.5 The shape of the electromagnetic emissions (EE) component in accordance with the IEC 61000-2-9, IEC 61000-2-10, IEC 61000-2-11 and MIL-STD-461 F standards.

6.3 The Parameters for Testing Immunity to Intentional Electromagnetic Interference (IEMI)

As has been noted above the IEC 61000-4-36 standard with the test parameters for testing the immunity to IEMI have not yet been published, however there are other standards as well as the results of research, that characterize the parameters of IEMI [6.3–6.15]. At the present time several different methods are in existence for generating high-power RFE, which can be used for attacking electronic and computer systems remotely, which give rise to a very broad range of emission parameters:

The electrical field strength ranging from 1 to 100 kV/m;
A pulse leading edge duration ranging from 100 to 500 ps;
A pulse duration ranging from hundreds of pinoseconds to units of nanoseconds;
Pulse repetition rate - from 01. - 1,000 GHz

It is completely obvious that given such a broad range of parameters for the sources that have been developed it is very difficult to establish any kind of clear requirements for testing electronic apparatus for immunity to these emissions. Nevertheless on the basis of research conducted by a leading specialist in this field William Radasky it is possible to speak of a broad band pulse emission with a leading edge duration of 100 ps, and a pulse width of 1 ns, and a pulse repetition rate of 1 MHz and an electrical field strength of 10 kV/m [6.7]. According to information the author has in his possession, these parameters should be entered into the IEC 61000-4-36 standard.

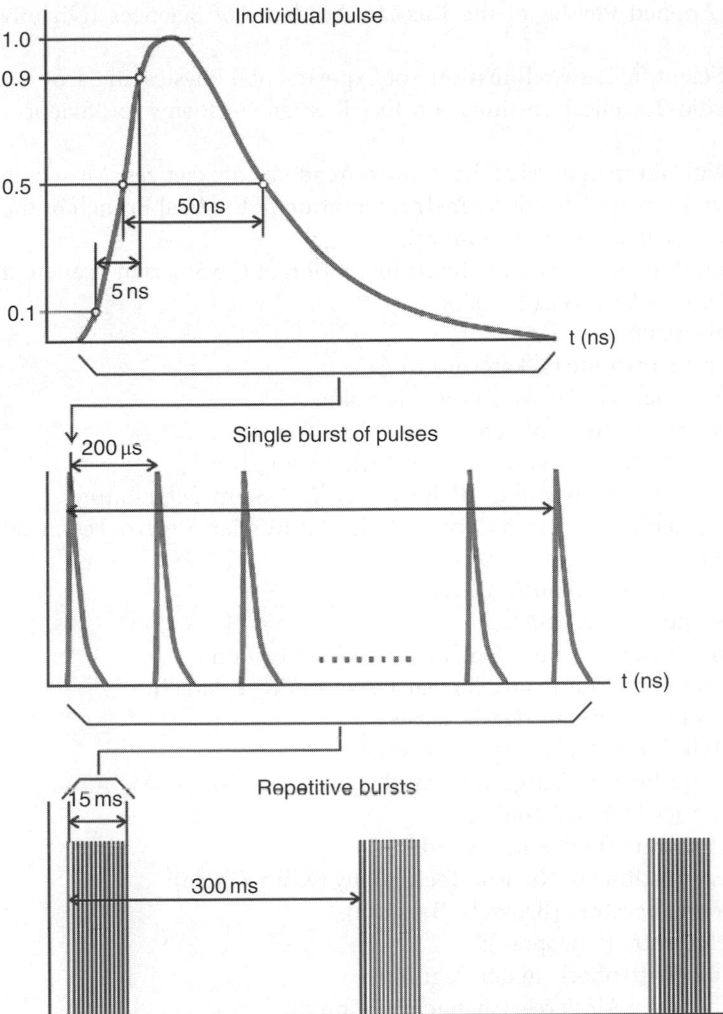

Fig. 6.6 Electrical Fast Transient (EFT) pulse (IEC 61000-4-4).

6.4 Testing Equipment for Testing Immunity to HPEM

A great number of organizations in many countries across the world are engaged in the development of sources of high-power pulses and radio-frequency electromagnetic emissions:

1) The Institute of High Current Electronics of the Siberian Branch of the Russian Academy of Sciences (Tomsk);
2) The Institute of Electrophysics of the Ural Branch of the Russian Academy of Sciences (Yekaterinburg);
3) The Institute of Radio-technology and Electronics of the Russian Academy of Sciences (Moscow);

4) The Institute of Applied Physics of the Russian Academy of Sciences (Nizhniy Novgorod);
5) The All-Russian Scientific Research Institute of Experimental Physics, in Sarov
6) The Moscow Radio-Technical Institute of the Russian Academy of Sciences (Moscow);
7) The Institute of High Temperatures of the Russian Academy of Sciences (Moscow);
8) The Special Design Bureau of Scientific Instrumentation of the Ural Branch of the Russian Academy of Sciences (Yekaterinburg);
9) The Special Design Bureau of Scientific Instrumentation of the Siberian Branch of the Russian Academy of Sciences (Tomsk);
10) Moscow State University;
11) The Urals Polytechnic Institute (Yekaterinburg);
12) The Production Association Tomsktransgas (Tomsk);
13) The Scientific Research Institute of Semiconductors (Tomsk);
14) The Scientific Production Association "ZENIT" (Zelenograd);
15) The Scientific Production Association "BUREVESTNIK" (Saint-Petersburg);
16) The High-Voltage Scientific Research Centre of The All-Russian Electro-Technical Institute (Moscow);
17) The Texas Tech University (Lubbock, USA);
18) Advanced Physics, Inc (Irvine, USA);
19) The GEC-Marconi Research Centre (Chelmsford, Great Britain);
20) The SOREQ Experimental Centre for Nuclear Physics (Yavne, Israel);
21) Rafael Advanced Defense Systems (Haifa, Israel);
22) The DSTO Research Centre (Salisbury, Australia);
23) The University of Strathclyde (Glasgow, Great Britain);
24) The Institute of Physics (Tartu, Estonia);
25) The FOA Research Centre (Linkoping, Sweden);
26) The North-Western Institute of Nuclear Technology (Xi'an, China);
27) The RMA Research Laboratory (Brussels, Belgium);
28) The DSO Research Centre (Singapore);
29) The Diehl Shtiftung (Rottenbach an der Pegnitz);
30) Aviation University of the Air Force, Changchun, China;
31) Electrostatic and Electromagnetic Protection Research Institute, China;
32) University of Electronic Science and Technology, China;
33) Beijing Key Laboratory of High Voltage & EMC, China;
34) North China Electric Power University, China;
35) Key Laboratory of Power Systems Protection and Dynamic Security Monitoring and Control, China;
36) Nanjing Engineering Institute No. 1, China;
37) Jilin University, China

However, production in the majority of these organizations is designed to meet the needs of their own research and is not destined for sale on the open market as electromagnetic emission or IEMI simulators with the aim of testing electronic apparatus. The majority of the large manufacturers of military technology have their own test rigs to test examples of their own technology, Fig. 6.7., which is also not available for sale.

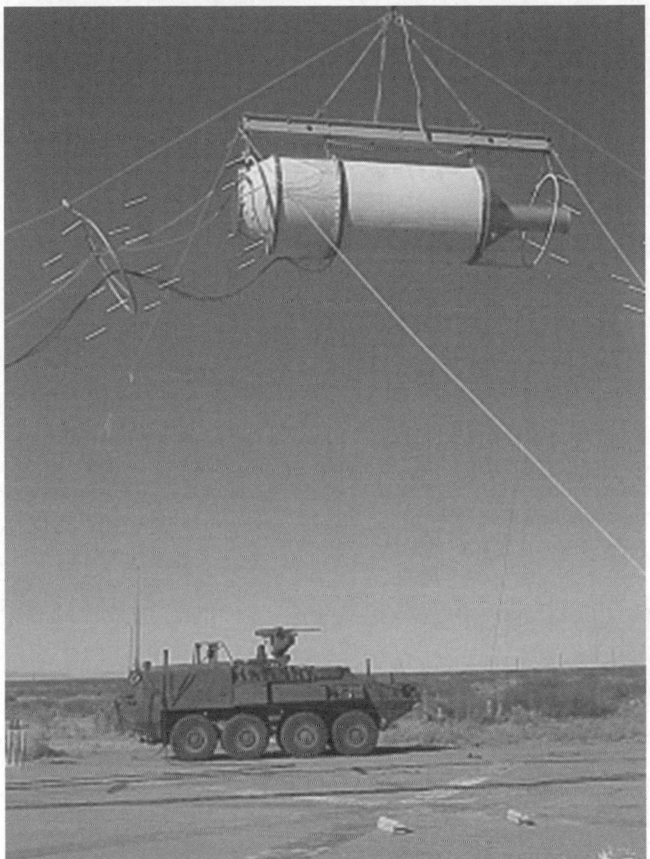

Fig. 6.7 A test rig designed to simulate the effect of HEMP on military technology.

Manufacturers of military technology in Canada, China, France, Germany, India, Israel, Italy, Holland, Russia, Sweden, Ukraine, Great Britain, and the United States also have these same rigs. Some of them are described in detail in the IEC/TR 61000-4-32 standard.

In many countries there are specialized testing laboratories that take orders for conducting tests of this nature from external organizations. In Russia this is represented by the test complex "Allure" at the High-Voltage Research Centre of the Federal State Unitary Enterprise The All Russian Electro-Technical Institute in Istra in the Moscow region, Fig. 6.8. There are however very few installations in the world for simulating electromagnetic emissions and the companies that are producing test equipment commercially for conducting tests on the resilience to HPEM in accordance with the standards are Dayton T. Brown (USA), and Aero-Rad Technology Co Ltd (China) amongst others. Small test installations which are suitable for laboratory testing of microprocessor based electronic relay type electronic apparatus for resilience to electromagnetic emissions are only produced by two companies: The Swiss department of Montena Technology and the American company Applied Physical Electronics, Fig. 6.9.

Fig. 6.8 A stationary "Allure" HEMP simulator at the High-Voltage Research Centre of the Federal State Unitary Enterprise The All Russian Electro-Technical Institute in Istra in the Moscow region. The dimensions of the simulator are as follows: 100 x 35 x 13.5 m; the operating volume is: 10 x 10 x 10 m, the shape of the pulse: 2.5/25 ns; the maximum field strength of the pulse electrical field: 70 kV/m.

(a)

(b)

Fig. 6.9 Compact rigs for testing electronic apparatus for immunity to electromagnetic emissions. (a) Montena Technology, (b) Applied Physical Electronics.

Fig. 6.10 A test rig produced commercially by Montena Technology (in the United States of America) for laboratory testing of larger objects, such as protective relay cabinets for example for immunity to HEMP. On the left a pulse generator is visible, while on the right is an antenna system.

Montena Technology produces many different types of test equipment including relatively large rigs (with a height of 1.8 m, and a length of 7 m), which are suitable for testing entire cabinets containing electronic apparatus, such as protective relay cabinets for example, Fig. 6.10.

As far as test equipment for testing electronic apparatus to IEMI is concerned then different types of compact Marx generators with different outputs can be used for this purpose, which are equipped with a unidirectional antenna, a wide range of which are produced by Applied Physical Electronics, Fig. 6.11.

HEMP has the following features that need to be considered as the test procedure is developed. The disturbance has a very short time length (a single pulse lasts for several nanoseconds); therefore, the failure of the equipment under test (EUT) must be registered within this period. This severely limits the number of EUT modes of operation to be controlled during the test. For example, during the HEMP immunity test, it is not possible to change the EUT modes of operation using the PC connected to the EUT in order to observe the EUT response to these changes (as is possible during the normal EMC test in the so-called anechoic chamber, where the EUT is exposed to electromagnetic radiation for a long period).

There is a danger that so-called soft faults, soft failures, or soft errors may appear especially in electronic memory elements, which are difficult to discover instantly during the tests. The soft faults may show up in the tested apparatus only after a long period, for example, on accessing the destroyed memory cells during certain operations, or certain program modules.

HEMP simulators usually consist of a concrete base with a bonded-in metal mesh acting as the first electrode, and the overhead metal mesh located 10–15 m above the concrete base, acting as the second electrode. A high-voltage pulse sent from the output

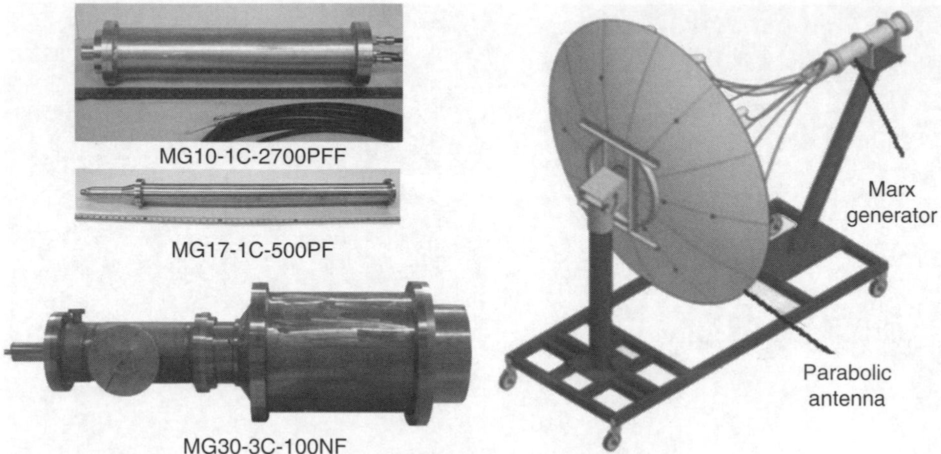

Fig. 6.11 Compact high-power Marx generators - MG10-1C-2700PFF (300 kV, 1 GW); MG17-1C-500PF (510 kV, 400 MW); MG30-3C-100NF (600 kV, 6 GW) and high-power radio-frequency electromagnetic emissions source based on these generators equipped with a unidirectional parabolic antenna.

of the special generator (usually, a Marx generator built on a set of high-voltage capacitors and triggered spark gaps) is applied between these two electrodes. Generally, the bonded-in mesh connected to the generator cannot be used as the EUT grounding system.

These tests are expensive due to the limited number of test centres in the country and their affiliation to the Ministry of Defense.

It is difficult to select the right EUT configuration, unlike the regular electromagnetic interferences, HEMP is more global than local in nature and affects both EUT and its feeding system, grounding system and channels of communication with other facilities. Therefore, instead of a single device, unit or module, a whole system of interconnected devices, units and modules, simulating the real-life environment, including a set of different and distanced grounding points, needs to be tested.

Due to the complexity and the high cost of the HEMP resilience tests, these tests should be applied only to a limited range of equipment types considered as critically important devices, the lack of which makes even partial operation of electric energy facilities impossible. The selection of the equipment should be the first stage of the test plan development. The second stage should include the clear and transparent description of the test objective, as it will define both the test object and the test procedure. Potential test objectives can be defined as follows:

1) Test the resilience of operating equipment to the highest possible HEMP level without any protection. The objective of this test is to discover elements and systems exposed to HEMP, and thus requiring protection.
2) Test the effectiveness of the equipment protection against the maximum possible HEMP level using the minimum set of pre-installed protective means designed for power system operation. These tests can be use to check the performance of the minimum set of protective means and to discover the types of equipment failures expected under HEMP.
3) Test the effectiveness of the equipment protection against the maximum possible HEMP level using the complete set of pre-installed protective means designed for

newly constructed power systems. This test allows the effectiveness of the most complex and expensive type of protection to be confirmed and justifies the costs of a protection system.

4) Test the operating equipment without special protective means under a series of pulses with the amplitude gradually increasing from 20% to 100% of the maximum possible level. The objective of this test is to (1) find the equipment (or equipment types) most exposed to HEMP and (2) define the maximum HEMP amplitude that the unprotected equipment can safely withstand, in order to calculate the required level of additional protection to improve the equipment's own attenuation to the maximum level provided by the standard.

As mentioned earlier, the number of controllable EUT operation modes during the test is very limited. Thus, these modes can include unauthorized appearance or disappearance of signals at outputs of at least two EUTs interconnected via communication channels in the standby mode (waiting mode), or to high data exchange mode (e.g. in emergency mode). In the latter case, the simulation of the emergency mode start should be synchronized with the initiation of the HEMP pulse.

After the test, because the complex microprocessor-based electronics may have unevident soft failures (even if no damage or failures that have appeared during the test are registered), the equipment should further undergo a complete and thorough functional check. This means that the test designed to determine the required level of additional protection should be followed by a functional check initiated after each level of HEMP impact. Clearly, this makes the test set much more complex, as after each higher amplitude pulse, the EUT should undergo a functional check. The EUT should be reconnected to the functional check systems after each HEMP test cycle. A programmable portable test system, pre-programmed to perform certain functional checks, can simplify the test process. Many companies manufacture such systems (DOBLE, ISA, Omicron, Megger, etc.), and they are widely used in relay protection testing.

Since under the effect of HEMP the grounding system acts as a huge antenna absorbing electromagnetic energy over a wide area and delivering it directly to grounded electronics, the test program must include ground system impact tests performed at two EUTs, at a distance from one another and connected to the same grounding system at two remote points. As the HEMP simulator is not permitted to use the mesh bonded in the concrete base as the EUT grounding system, a separate grounding system, which is realized in terms of a sufficiently large mesh, must be provided for the purposes of the test. The presence of both horizontal and vertical HEMP components means that such a grounding mesh should be installed at 30–45° to the concrete base rather than horizontally. The mesh can be assembled from separate sections connected together using a corresponding wire.

Generally speaking, power system electronic devices are connected to other electronics, sensors, power sources, electrical or electromechanical power units combined into a complex environment. For example, the relay protection system, SCADA (Supervisory Control And Data Acquisition system), a fire protection system, is built on such a principle. Thus, the tests should cover the complete system rather than a single unit. For relay protection systems, such an arrangement can include two cabinets each equipped with a digital protective relay (DPR), the battery acting as a power source and a battery charger. These cabinets should be spaced apart to the maximum allowable distance, connected to the grounding mesh and interconnected via a communication channel. DPR current and voltage inputs should be connected to the controlled current and voltage sources protected from the test pulse impact. To arrange such protection, the source should be located in the isolated shielded section and connected to the EUT via a special filter (coupling–decoupling circuit), allowing the transmission of signals between the EUT and the protected equipment placed in the isolated section, but preventing the ingress of the HEMP test pulse. The current and voltage source must be fitted with remote controls to synchronize its start-up with the HEMP test pulsing cycle. The state of DPR output circuits should also be considered during the test, see Fig. 6.13. Usually, stationary test beds are pre-equipped with special shielded cables and filters designed for data transfer from the EUT, installed onto the test bed to the isolated shielded section.

As a rule, the power system electronic equipment is installed in metal cabinets located in all-brick or concrete buildings that significantly attenuate the impact of HEMP, while other devices, such as the grounding system, sensors, measuring current and voltage transformers, and numerous outgoing cables, are located in an exposed area. Therefore, in real life, the different components of the common system are exposed to the electromagnetic impacts of different energy. The traditional design of the test-bed radiating antenna, Fig. 6.12, includes the central section, where two parallel meshes

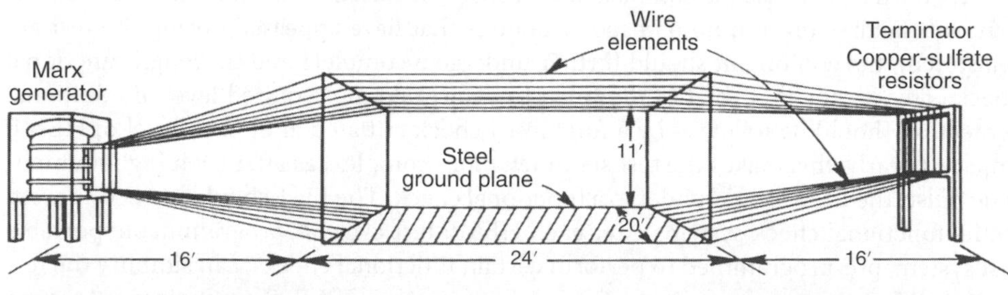

Fig. 6.12 The design and appearance of the HEMP simulator.

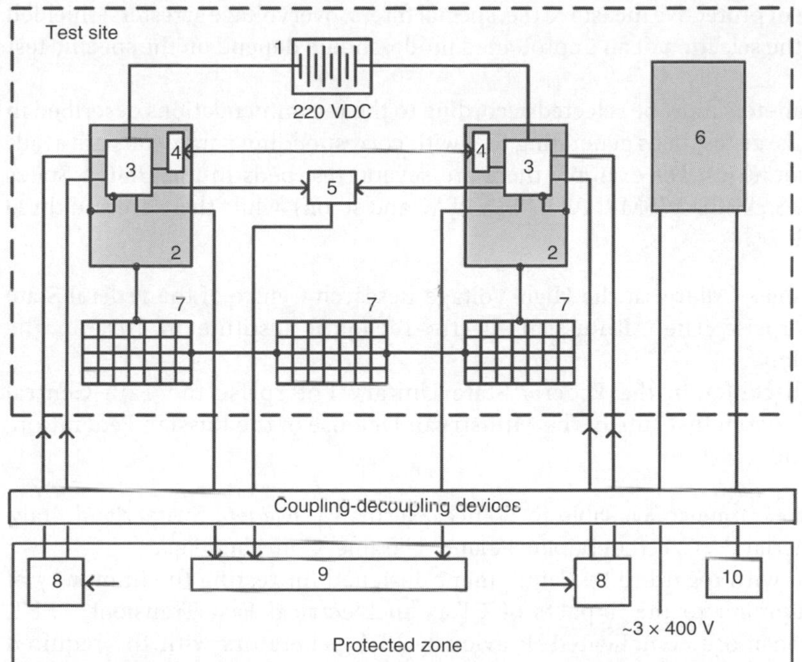

Fig. 6.13 Arrangement of EUT in the test bed. 1 – Mobile battery 220V; 2 – electrical cabinets distanced from one another; 3 – tested electronics (i.e. digital protective relays or DPRs); 4 – communication devices; 5 – lockout relay controlled via DPR output circuits; 6 – battery charger; 7 – set of metal meshes comprising the ground system model; 8 – simulators of different modes of EUT operation synchronized with HEMP initiation system; 9 – EUT status recorders; and 10 – load with battery charger output voltage control unit.

are located at a fixed distance to each other, and two side sections, where the distance between the top and bottom meshes decreases. Such design allows different components of the system to be tested under the varied impacts of the same test pulse, since the strength of the field between the top and the bottom meshes depends heavily on the distance between them. To create the conditions most relevant to real-life situations, the components of the tested system should be located on the different sections of the test bed.

The presence of both the vertical (directed from the high-altitude detonation point to the ground surface) and the large horizontal elements of the E1 component of HEMP is another aspect which needs to be considered during the HEMP test. To allow for the consideration of both HEMP components, the elements of the system being tested, which are located in between the top and the bottom meshes should be arranged at a certain angle to the ground surface.

The test procedure should include the registration of the EUT parameters under the HEMP using proper and adequately protected recording equipment. These could include external loggers and memory impulse scopes with automatic triggers as well as emergency event recorders built in the EUT and operating in parallel with other recording devices.

The selection of protective measures (i.e. special filters, overvoltage arresters, shielded cables, etc.) or the selection of an unprotected mode, should depend on the specific test objective.

The test parameters must be selected according to the recommendations described in paragraph 6.2. Large test beds generating EUI with corresponding parameters are available in many countries. For example there are several test beds in the United States (TORUS, ALECS, ARES, WSMR, ATLAS, VPBW and so on), while there are and three in Russia as follows:

- The test complex "Allure" at the High-Voltage Research Centre of the Federal State Unitary Enterprise The All-Russian Electro-Technical Institute in Istra in the Moscow Region
- The research center in the Federal State Unitary Enterprise the 12th Central Scientific-Research Institute of the Ministry of Defense of the Russian Federation, Sergiyev Posad

Such simulators are also available in France, Germany, Sweden, Switzerland, Italy, Israel, the Netherlands, Czech Republic, Poland, Ukraine, China and Japan.

The situation with regard to test equipment designed for testing the immunity of electronic equipment to the impacts of CI as an Electrical Fast Transient - EFT, Fig. 6.12 is even more complicated. Previously EFT generators with the required level of peak output voltage of 8 kV were produced by TESEQ, Kentech Instruments Ltd and Thermo Electron Corp (Table 6.1.) and were based on a vacuum triggered spark gap that would form test pulses. With the advent of high-power semiconductor switching elements - IGBT -transistors the production of vacuum triggered spark gaps was stopped by all three companies, since the pulses formed by the transistors turned out to be slightly more stable and "correct" than pulses formed by the vacuum spark gaps. Unfortunately along with an enhanced stability in the generated pulses, their amplitude was reduced.

An analysis that we carried out demonstrated that today not one of the commercially produced EFT generators fully satisfies the requirements of the standards for pulse amplitude (8 kV). The closest to these requirements for the pulse amplitude value is the PEFT 8010 type generator produced by the Swedish company Haefely EMC Technology, Fig. 6.14.

Thus for testing DPR for resilience to HPEM three types of impacts are required, that are produced in addition to the full package of testing for electromagnetic compatibility [6.17];

1) A pulse electromagnetic emission with a leading edge pulse duration of 2 ns, a pulse width of 25 ns and with a field intensity of 5-50 kV/m;
2) A pulse electromagnetic emission with a leading edge pulse duration of 100 ps, a pulse width of 1 ns, a pulse repetition rate of 1 Mhz and a field intensity of 10 kV/m;
3) A fast pulse of 5/50 ns (EFT) with a pulse amplitude of 8 kV, delivered via the contacts to the inputs in a DPR.

A compact test apparatus with parameters that are sufficiently close to the requirements is freely available on the market. This makes the organization of a specialized

Table 6.1 The peak voltages for of 5/50 ns pulse for EFT generators (IEC 61000-4-4) that are available on the market.

Type of EFT Generator	Manufacturer	The peak voltage of output pulse, kV
PEFT 8010	Haefely EMC Technology	7.3
NSG 2025*	TESEQ	8
J0101031/3*	Kentech Instruments Ltd.	8
KeyTek ECAT E421*	Thermo Electron Corp.	8
FNS-AX3-A16B	NoiseKen Laboratory Co.	4.8
EFT 500 N8	EMTEST	7
TRA3000	EMC Partner	5
EFT 6501	Schaffner	4.4
EFT-4060B	Shanghai Yi PaiElectronmagneticTechn.	6.6
EFT500	Suzhou 3Ctest Electronic Co.	5
AXOS8	Hipotronics	5

* Production has halted

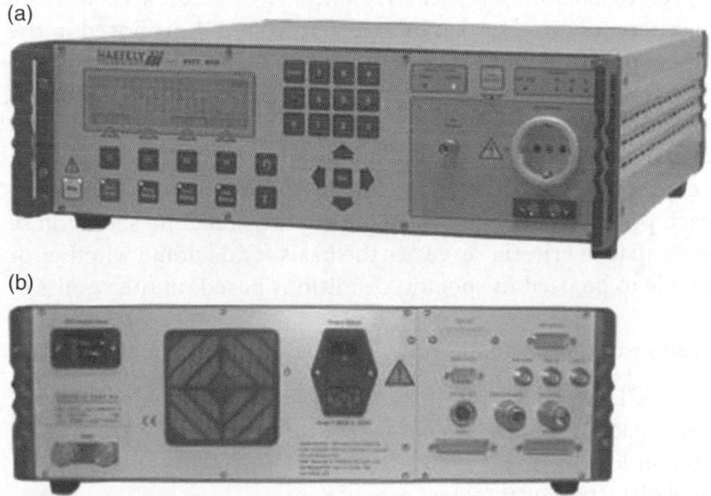

Fig. 6.14 A PEFT 8010 type EFT generator with a maximum pulse amplitude of 7.3 kV, produced by Haefely EMC Technology (in Sweden), (a) A view of the front panel, (b) A view of the rear panel.

laboratory entirely possible and accessible, which could be used to check the immunity of modern relay protection devices as well as other types of so-called "critical" industrial electronic apparatus for resilience to HPEM. The cost of a complete set of equipment for such a laboratory would be around $500,000.

6.5 Use of the Performance Criteria During Testing of Electronic Apparatus for Electromagnetic Compatibility (EMC)

The reaction of an object undergoing testing to electromagnetic impacts could be different. For example the object may fail completely due to a flashover in the electronic components, or perhaps it could temporarily loose its serviceability just for the duration of the impact of a pulse or an electromagnetic field. Another variant is that there may be a short failure in the software under the influence of a pulse voltage that has been applied to it, which either requires or does not require a subsequent reboot of the internal software by an operator conducting the test. There may be a number of different variants of the reactions of the object undergoing testing to electromagnetic impacts. The acceptable reaction of an object undergoing testing to electromagnetic impacts for a given type of object or a given type of test is known as the "performance criteria." These performance criteria are the most important indicator during testing for EMC since the conclusion of whether a given device has passed the test successfully or not is dependent on selecting these correctly. However there are no methods, nor could there ever be, any methods by which these can be selected correctly. As a rule, everything is limited by a phrase such as this: "*The selection of the degree of hardiness, and of the performance criteria is taken by those who develop, agree, and approve the technical specifications or technical regulations*" and by a table from which one performance criteria or another can be selected from 3-4 proposed in the specific standard. This is understandable, since the right selection depends on the specific object undergoing testing and on the specific modes and conditions in which it operates. Furthermore different performance criteria can be selected for one and the same object dependent on its specific operating mode, connection, the function it performs, its operating conditions and so on. Therefore an understanding of the specific idiosyncrasies of each particular object is very important, since the selection of one or another of the performance criteria serves as the basis for deciding whether or not a given object is suitable to be used in specific conditions based on the results of the test.

Common acceptable responses (performance criteria) for EUT are the following:

- Graphical distortions on EUT display, display flickers or blinks off.
- Display showing incorrect data
- Signals or data distortion or loss
- Communication channel distortion or total loss
- Sensor malfunctions
- System false activation
- Sharp degradation of the system's ability to handle and transmit the information, and maloperation of the system
- Software failures
- The system hanging
- An automatic restart of the computerizing electronic system
- Complete system failure due to a power source fault or a fuse blowing in feed circuits
- Physical damage to the internal electronic components of EUT

Basic standard IEC 61000-4-25 (Part 9) describes only the following five types of performance criteria summarizing the EUT responses listed earlier:

1) Normal performance within specified limits
2) Temporary degradation or loss of functions or performance which is self-recoverable
3) Temporary degradation or loss of functions or performance, which requires operator intervention or system reset
4) Degradation or loss of functions which is not recoverable, due to a loss of data or damage to equipment
5) Degradation that can lead to a safety problem (e.g. fire)

For the test planning, the performance criteria should be predetermined separately for each type of test. These criteria should be used for assessing the EUT test results for each type of test. Apparently, only criteria 1 and 2 are relevant to power system electronics, so these are the only two acceptable choices.

6.6 The Idiosyncrasies of using Performance Criteria during Testing of Microprocessor Based Relay Protection Devices for their Immunity to HPEM

The organization NERC (North American Electric Reliability Council) at the request of a special commission under the U.S. Congress 'The Commission to Assess The Threat to the United States from Electromagnetic Pulse (EMP) Attack" drew up a list of power engineering equipment, which should be tested for its immunity to the impact of HEMP. This list included, specifically, microprocessor based protective relays and the SCADA system devices (Supervisory Control and Data Acquisition) - the collective name for the hardware and software apparatus of different types that ensure the collection of different types of data in real time from a number of sensors, as well as ensuring the processing, archiving, display, and also the transfer of information concerning the objects undergoing monitoring, as well as sending the operator's commands to remote objects - this serves as the basis for the modern Automated Process Control Systems in substations. Metatech Corp. (USA) conducted tests on SEL-311L type DPR (differential line protection) and an SEL-2032 type controller, Fig. 6.15. in an accelerated test program and solely to test immunity to the E1 component of HEMP. The results of these tests are presented in a Meta-R-320 report [6.16]. As this report shows an assessment of the serviceability and of the lack of damage ***after each test*** were set as the performance criteria during the DPR and the SCADA controller tests, linked to the feed of short (5/50 ns) high-power pulses with an amplitude of up to 8 kV to various device inputs. As the report mentions, as the pulses were applied to successive ports with an amplitude of 3.2 kV, the DPR was activated inadvertently, but then returned to its normal operating mode. Several other ports (such as the IRIG - Inter-Range Instrumentation Group time code - the timing port) were damaged at a voltage of 600 V. In the SCADA controller the Ethernet communication module was damaged at a voltage of 1.2 kV.

Fig. 6.15 An SEL-311 L type DPR and a SCADA system controller produced by Schweitzer Engineering Laboratories (in the USA) and which underwent testing for their immunity to HEMP.

The report notes that one of the additional performance criteria included internal recording of the results of the oscillography of the currents and voltages applied to the relay inputs. The report notes that no violations were encountered in the testing process.

6.7 A Critique of the Method of Testing of the DPR Used in [6.16]

1) In our opinion the use of performance criteria based on the serviceability of the DPR ***after the impact of interference*** is not correct and does not enable an unambiguous conclusion to be made concerning the immunity of this DPR to this interference. This is linked to the fact that the DPR has certain specific idiosyncrasies in comparison to the SCADA system, which are examined in [6.17 & 6.18]. Given all the importance and the responsibilities of the SCADA system it is first and foremost designed to collect, process, and display information automatically. Despite the fact that Remote Terminal Units (RTUs) form part of the system they are not able to operate in an automatic mode and are designed purely to perform the commands given to them by an operator from a remote dispatch point. The majority of the modern power substations operate in an automatic mode without human intervention. Manual control of the position of circuit breakers in substations (which in actual fact is to manually control the configuration of the electrical network) is carried out by an operator from the dispatch centre using the SCADA system, which is characterized by its vulnerability to the impact of HPEM. Therefore in the event of the impact of HPEM remote control of the substation from the dispatch point would in all probability be lost and the configuration of the electrical network would be defined by the relay protection system alone - this

is the only system, which is capable of automatically influencing the position of the circuit breakers. Furthermore a DPR, which forms the basis of modern relay protection, is constantly sharing information and commands ***in an automatic mode along communication channels that are vulnerable to HPEM*** (in contrast to the SCADA system, in which the critical control commands are only sent to the circuit breakers on the initiative of the dispatcher). In the event of an automatically functioning protective relay failing, and the dispatcher not being able to intervene in its operation, specifically in the event of a spurious activation of circuit breakers under the influence of HPEM, the electrical network, and beyond that the energy system as a whole could be completely destroyed. This is one of the reasons why DPR should be tested for their immunity to the impact of HPEM ***whilst in operation*** and should not be tested for damage **after** the impact of interference on the relay.

2) The ways that electromagnetic interference can penetrate DPR in the form of pulses which for the most part are fed to unprotected inputs and in the form of a high-frequency electromagnetic wave that penetrates directly into the internal sensitive electronic components or via the unprotected inputs/outputs of electronic modules, or via the many cables connected to a DPR and which act as antennae, absorbing electromagnetic energy - are different. Furthermore the HPEM is not limited to HEMP but also includes ultra broad band high-frequency interference from specific sources with an output of several Gigawatts, intended to attack electronic apparatus remotely [6.18]. Unfortunately this threat is not only specially designed to attack the electronics of a piece of apparatus but also the emissions from standard high-power radars. Thus in 1999 for example a case was officially recorded of a catastrophic failure in the SCADA system used by the San Diego County Water Authority that supplied the city of San Diego with water using emissions from a ship's radar that was located 25 miles from the city. A similar case occurred in 1980 in Holland in a gas pipeline located some one and a half kilometres from the Port of Den Helder. On that occasion damage to the SCADA system using a port radar led to a powerful gas explosion. Therefore testing DPR for their immunity should not just be limited to applying high-power pulses to specific inputs, but should also be accompanied by exposing the object to electromagnetic emissions from a unidirectional antenna, which is envisaged in the corresponding standards.

It is worth considering that in the event of an HEMP attack its impact would be felt not only by electronic equipment (DPR, and SCADA system apparatus) but also in the power equipment within energy systems: linear insulators, transformers, and generators. Moreover it is not only the E1 component of HEMP (which was simulated in the [6.26] tests) but also the two other components: E2 and E3. From research conducted previously [6.16] in the Soviet Union and the USA it is known that as a result of the combined impact of all the components of HEMP the probability of damage to high-voltage equipment is very high: a failure in linear insulators, high-power transformers becoming saturated and burning out, generator insulator failures and so on. That is to say the moment at which the high-power electromagnetic interference acts on a DPR coincides with the moment the internal functional mode of the DPR changes, and is linked to the advent in its inputs of emergency levels for the currents and voltages that are being controlled. Is it possible for a protective relay that is being subjected to the impact of HPEM to disconnect a transformer that is approaching saturation, as well as the damaged section of an overhead line, or a broken cable? Would a simultaneous mass failures of different DPR cause a complete failure and a collapse of the power system?

Tests carried out in [6.16] do not provide any answers to these questions.

"We constructed such complex systems that we cannot envisage all the possible interactions within them, all the possible failures. We add all the new safety devices to these systems, but remain cheated and beaten by the hidden links within them" - wrote the famous specialist in the reliability and vulnerability of complex systems Charles Perrow. Charles called this problem and "an inconceivability" since even the most run of the mill failure initiates interactions, which are "Not only unexpected, but also unpredictable" for a certain critical period of time. In the majority of failures nobody had anticipated how one "interaction algorithm" would affect another, and as such nobody could predict what would happen in good time. This relates very well to the modern highly complex and multifaceted world of relay protection, in which the behavior of a protective relay under the influence of HPEM cannot be envisaged.

6.8 An Analysis of the Results of the Second Independent Test of a DPR of the Same Type

Another of the tests on a DPR of the same type (by a strange coincidence) is covered in a promotional presentation by the manufacturer of the equipment Schweitzer Engineering Laboratories [6.20] in which the results of tests on examples of SEL-311 L DPR conducted on test rigs at the US Army's Picatinny Arsenal in New Jersey during tests for HEMP and for electromagnetic emissions, Fig. 6.16. In this promotional presentation it is confirmed that all the tests were successful. Furthermore a closer analysis of this material revealed several oddities. For example, the advertisement depicted in Fig. 6.17. confirms that the SEL-311 L was tested under a field strength of within a range of 25 - 1,000 V/m, while the MIL-STD-461 military standard requires just 50 V/m.

Specialists at such a serious company as SEL demonstrate quite a strange lack of awareness in this document, if the fact that the MIL-STD-461 stipulates a field strength that corresponds to the HPEM are examined not in Volts but in kilovolts and the figure "50" is written not as 50 V/m but a 50 kV/m is taken into account.

The bar graph illustrated in Fig. 6.18 appears even more curious and it shows that a field strength of 1,000 V/m was indeed used during testing at frequencies of just 1,000 - 1,500 MHz and in the rest of the frequency range the field strength was a little over double this figure, while the amplitude to frequency response ratio does not correspond to the MIL-STD-461 standard.

As is evident from the graph the field strength levels in the chart are limited by the onset of instability in the performance of the relay (the yellow areas at the top of the bars). That is to say in actual fact this diagram illustrates an area of stable operation for the SEL terminal that is installed separately (that is to say outside of the protective relay system).

Fig. 6.16 A test of the SEL 311 L type DPR on the impact of HPEM on test rigs at the US Army's Picatinny Arsenal in New Jersey [6.20].

It follows from this that the relay does not provide a stable performance outside of the field of values set out in this diagram with its extraordinarily low values for the electromagnetic field strength. If this were compared to the aforementioned MIL-STD-461 standard, Fig. 6.19, it can be noted that the parameters that have been applied for the test impacts do not even come close to the requirements of this standard.

Considering this oddity in selecting the parameters for testing the SEL-311 for resilience to HPEM is it possible to relate to the manufacturer's confirmation that their relays are immune to HPEM?

Another problem is linked to the selection of the quality of the object undergoing testing in terms of a single DPR terminal. These terminals are as a rule manufactured in metal casings, which effectively attenuate the electromagnetic emissions, and as such it is wholly to be expected that the results of tests to assess the immunity to the impacts of electromagnetic emissions on a single terminal would be positive. In actual operating conditions there are a great number of cables that are attached to a microprocessor

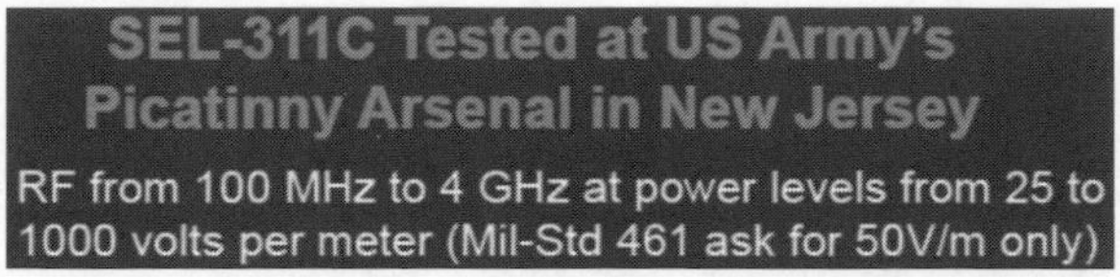

SEL-311C Tested at US Army's Picatinny Arsenal in New Jersey

RF from 100 MHz to 4 GHz at power levels from 25 to 1000 volts per meter (Mil-Std 461 ask for 50V/m only)

Fig. 6.17 The text taken from the SEL company advertisement [6].

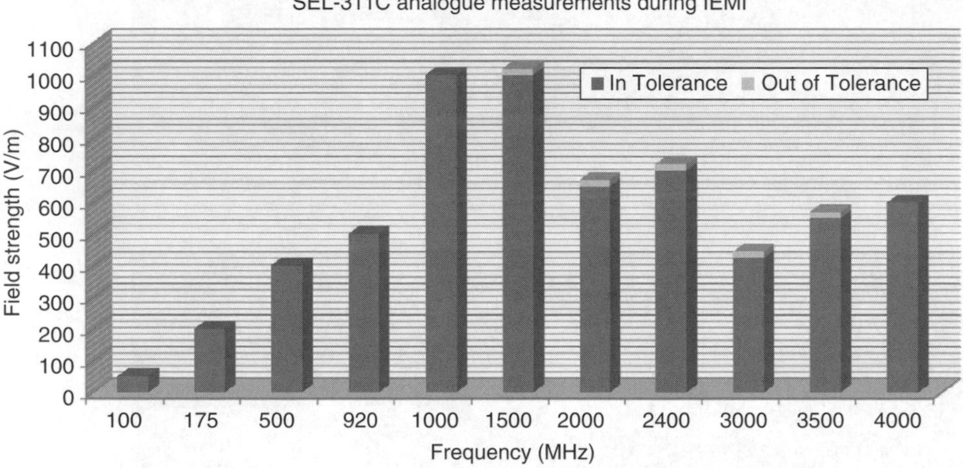

Fig. 6.18 The parameters of electromagnetic emissions during testing with the SEL-311 L microprocessor based protective relay [6.20].

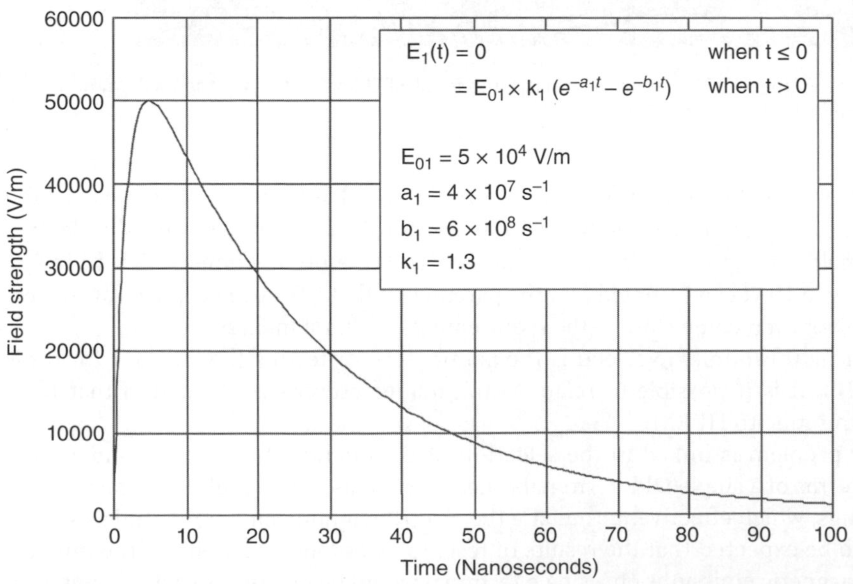

Fig. 6.19 A graph taken from p. 138 of the MIL-STD-461 standard for comparison with the graph illustrated in Fig. 6.18. (1 ns corresponds to a frequency of 1 GHz).

Fig. 6.20 Testing of the SCADA system for immunity to HEMP [6.21]. The antenna for the EMP simulation system is visible above. The elements of the SCADA system are located in separate boxes and are linked together by a standard communication system.

based protective relay that act as an antenna, absorbing electromagnetic energy and delivering it to the internal elements of the DPR; many of the DPR terminals are connected together via corresponding communications links that have been subject to the influence of HPEM. Therefore a relay protection system should undergo testing but whilst it is in operation and not as a separate terminal.

Testing of the SCADA system could serve as an example of the correct approach to testing complex systems, to which without a doubt relay protection belongs, described in [6.21], Fig. 6.20.

Thus it turns out that with the results of the two independent tests for the same type of microprocessor based protective relay conducted in two different ways it is not possible to reach a conclusion concerning its actual immunity to HPEM. Who actually needs these tests?

6.9 Conclusions and Recommendations for Testing Microprocessor Based Protective Relays

1) Due to the methodological mistakes during testing of DPR conducted previously by independent organisations they can never be acknowledged as satisfactory, but the results are significant. At the present time there is no reliable data on the degree of immunity of DPR to HPEM and as such these tests should be conducted again.
2) As a performance standard criteria should be selected that enable the performance of a DPR functioning in the normal and emergency modes for the object undergoing testing to be controlled as the electromagnetic interference is acting on it, and not criteria which are based solely on checking the serviceability of the DPR after the interference has ended.
3) Due to the complexity and the high cost of the HEMP immunity test, such a test should be applied only to a limited range of equipment types considered as critically important devices, the lack of which makes even partial operation of electric energy facilities impossible.

4) The HEMP immunity test plan should start from the clear and transparent definition of the test objective or objectives.
5) The power system electronics should be tested as a whole system rather than as a set of individual devices. A whole system such as this should include several electronic devices (two at least) interconnected by means of a communication channel, and connected to the common ground system, feeding source, control signal sources and so on. The plan development phase should include the charting of the system flow sheet and the compiling of a list of necessary equipment to perform the test.
6) Depending on the EUT type, the following should be defined in advance: (1) the list of parameters to be controlled over the HEMP impact period, (2) the parameter checkout methods and (3) the types of apparatus needed to record the changes of the parameters during the test performance.
7) The results of the HEMP impact can become evident only after a certain period following the test. Thus, the control of the EUT state during the test should be supplemented by the full EUT functional check performed after the test on the HEMP simulator test bed and then after feeding a high-voltage test pulse to the EUT using a contact method.
8) In addition to the complete set of standard EMC tests, the test for power system equipment immunity to the HEMP should take the following two types of impacts into consideration:
   i) Electromagnetic pulse – 2 ns rise time, 25 ns pulse width and 5–50 kV/m field strength
   ii) Fast pulse (EFT) fed to EUT inputs using contact methods – 5/50 ns, 8 kV amplitude
9) Standard criteria A and B should be selected as a performance criterion for power system electronics.

References

6.1 Platt R., Anderson B., Christofferson J., Enns J., Haworth M., Metz J., Pelletier P., Rupp R., Voss D. Low-frequency multigigawatt microwave pulses generated by a virtual cathode oscillator. - *Applied Physics Letters* 27.03. 1989, Vol. 54 Issue 13, p. 1215.

6.2 Counter-Electronics High Power Microwave Advanced Missile Project (CHAMP) Joint Capability Technology Demonstration (JCTD). Solicitation Number: BAA-08-RD-04, 16 October 2008 (Restricted Data).

6.3 Methods of Ensuring the Resilience of Future Radio-Relay, Tropospheric, and Satellite Communications to the Impact of High-Power Electromagnetic Interference / Voskobovich Vladimir Vikorovich 05.12.13 Moscow, 2002.

6.4 Development of the Methods of Assessing the Resilience of Telecommunication Systems to the Impact of Ultra Broad Band Electromagnetic Pulses / Vedmidskiy Aleksandr Aleksandrovich 05.12.13 Moscow, 2003

6.5 Theoretical and Experimental Methods of Assessing the Resilience of Terminals to the Impact of Ultra Broad Band Electromagnetic Pulses / Akbashev Beslan Borisovich 05.12.13 Moscow, 2005.

6.6 Methods and Techniques for Assessing the Impact of an High-Energy Electromagnetic Pulse on Telecommunications Networks / Yakushin Sergey Pavlovich 05.12.13 Moscow 2004.

6.7 Radasky W, Savage E, Intentional Electromagnetic Interference (IEMI) and Its Impact on the US Power Grid - Meta-R-323 Metatech report for Oak Ridge National Laboratory, 2010.

6.8 NATO AECTP-250 Leaflet 257 - *High Power Microwave* (HPM).

6.9 MIL-STD-461 F Requirements for the Control of Electromagnetic Interference Characteristics of Subsystems and Equipment, 2007

6.10 Staines G Compact Sources for Tactical RF Weapons Application - *DIEHL Munitionssyteme Amerem-2002 Maastricht,* Nederlands, 2002

6.11 R. Barker and E. Shamiloglu *High Power Microwave Sources and Technologies.* IEEE Press, New York, 2001.

6.12 W. Prather, C. Baum et al Ultra-wideband Source and Antenna Research *IEEE Trans. Plasma Sci.* Vol. 28 pp. 1624-1630, Oct. 2000.

6.13 Multi-wave High-Power UHF Generators / Koshelev V.I. 01.04.04 Tomsk, 1990

6.14 High-Power Pulse UHF Generators Based on a Backward Wave Tube in an Amplified Emissions Mode / Yelchaninov A.A. 01.04.04 Tomsk, 1990

6.15 The Generation of High-Power UHF emissions Based on High-Current Nanosecond Electronic Beams / Korovin, S.D. 01.04.04 Tomsk, 1990.

6.16 Savage E., Gilbert J, Radasky W. The Early-Time (E1) High Altitude Electromagnetic Pulse (HEMP) and its Impact on the U.S. Power Grid - Report Meta-R-320 for Oak Ridge National Laboratory, 2010

6.17 Gurevich V.I. Problems of Standardisation in the Field of Microprocessor Based Protective Relays - *Components and Technology,* 2012, No. 1, pp. 6–9.

6.18 Gurevich V.I. The Vulnerability of Microprocessor Based Protective Relays. Problems and Solutions - M.: *Infra-Inzheneriya,* 2014, 256 p.

6.19 Perrow C. *Normal accidents. Living with high risk technologies,* First ed. Princeton: Princeton University Press, 1984

6.20 *EMP Effects on Protection and Control Systems* - Schweitzer Engineering Laboratories, 2014, 31 p.

6.21 Report of the Commission to Assess the Threat to the United States from Electromagnetic Pulse (Attack), April 2008

7

Administrative and Technical Measures to Protect DPR from EMP

7.1 Problems with the Standardization of DPR

The standardization and universalization of DPR is one of the most important directions in enhancing the survivability of relay protection under the influence of EMP, since it enables an acceleration of the process of restoring the serviceability of relay protection after damage as a result of an electromagnetic pulse or another kind of remote electromagnetic attack and is also one of the ways in which the efficiency in the exploitation of a DPR can be enhanced in normal operating conditions.

7.1.1 Who Coordinates the Process of Standardization in the Field of Relay Protection?

Today, DPR devices are produced by dozens of the largest global companies such as ABB, Siemens, General Electric, Alstom (Areva), SEL, Nari Relays, Beckwith Electric, Schneider Electric, Cooper Power, Orion Itali, VAMP and Woodward among others, as well as a number of companies in Russia and Ukraine (ABB Rele-Cheboksary, The Scientific Production Association Ekra, the Scientific Production Association Bresler, the Closed Joint Stock Company the Cheboksary Electrical Equipment Factory, Radius-Avtomatika, Khartron-Unkor, Kievpribor, RELSiS, RZA Systems, Energomashvin, The Scientific Technical Centre Mekhatronika and so on.

Each of these manufacturers of DPR sets the dimensions and shape for each of the models they produce, as well as the composition and design of the internal modules and the internal programme interfaces. As a result, there are hundreds of different DPR on the market today, which are absolutely incompatible with one another both in terms of software or hardware. Moreover, in the majority of cases the different models or different generations of DPR produced by the same manufacturer are it appears incompatible with one another.

This situation leads to serious problems, which slow down the development of relay protection. Thus the incompatibility of DPR hardware leads to a situation where the consumer, in making a one-off purchase of an expensive protective relay package, is forced to acquire spare modules over the course of several years, which are far from cheap and that can only be purchased from the same manufacturer, even if the quality of their products leaves something to be desired. The acquisition of relay protection by

Protection of Substation Critical Equipment Against Intentional Electromagnetic Threats, First Edition. Vladimir Gurevich.

tender, which is happening in many countries across the world now, gives rise to a number of problems. The fact that manufacturing companies are victorious in tenders that are issued periodically leads to a situation in which, over the last 10–15 years, dozens of very different DPR modules produced by different manufacturers have accumulated in a single energy system, which sharply increases the workload for operations personnel and increases the number of mistakes made at every stage: from calculating the settings to introducing these settings and testing the relay. According to data that has been published by various authors the percentage for failures in DPR that are caused by human error (the so-called 'human factor' is up to 50–70% [7.1–7.3]. The different programme interfaces coupled with the complete incompatibility of the DPR programme does not enable an automation of the integrated testing process for DPR, which today requires a great deal of time and effort on the part of operational personnel, and which represent another opportunity for human error (e.g. re-setting the relay to its default settings following a test) [7.4].

Linked to the lack of standards, that is to say, in actual fact the absence of a framework the modern trends in the development of DPR continue to make them ever more complicated, and the number of functions built into them increases, and DPR continue to be used for tasks that do not bear any relation to relay protection, such as monitoring the condition of electrical equipment, which in our opinion is completely unacceptable given that many specialists are calling for the degree of autonomy for DPR to be enhanced [7.5–7.12]. It is worth noting that the trend for making DPR ever more complicated is one that is observed across the world and almost all the DPR manufacturers are guilty of this. In fact, the more complicated and more 'gimmicky' devices are easier to advertise and can be sold for a higher price. The consumers, however, do not assess this trend in relay protection in the same way as the manufacturers. This, for example, is how Russian relay protection specialists assess a protective relay produced by some of the leading DPR manufacturers [7.13]:

> *An untested technical and informational redundancy has been built into the SIPROTEC 7SJ642 (Siemens) terminal. In the manual (C53000G1140C1476, 2005) it is stated that 'The device is simple to use either via the integrated control panel or by attaching the PC to a DIGSI system programme', which does not correspond to reality. For example, it requires the integration of around 500 parameters (settings) that influence the operation of the device (the printed DIGSI signal matrix runs to around 100 pages of English text). Taking into account the need to formulate specifications for set up and protocols for testing the device, which need to set out all the generic parameters, the amount of documentation becomes unworkable. The enormous volume of input data makes setting up the device more complicated. The informational redundancy increases the risk of mistakes linked to the human factor. The technical redundancy requires the terminal operators to be highly qualified. The manufacturer's documentation on the terminals that have undergone examination runs into thousands of pages, but even then the required information is not there, and mistakes can be found.*

Unfortunately, this situation has existed in the DPR market for many years and continues to gather pace and the organizations that are tasked with coordinating the activities in this field (such as the All-Russian Scientific Research Institute for Relay

Production in Russia for example) are not only failing to find a solution to this problem but are actually supporting the exacerbation of this situation, by proposing that the functions of a DPR be expanded to an unbelievable degree by hanging functions off them, which bear absolutely no relation to relay protection, and using a non-deterministic logic as well as so-called 'artificial intelligence'. Thus, specifically, it is proposed that DPR be given functions for monitoring high-power electrical equipment and for forecasting its condition. It is proposed that a protective relay of this nature would be able to disconnect electrical equipment based on the results of an assessment of its own forecasts, long before the onset of an emergency mode (a so-called 'preventative action relay' [7.10–7.12]). This begs the question what exactly is a 'protective relay' if it can make forecasts and disconnect electrical equipment prior to the onset of an emergency mode.

Today no legitimate definition, even of such a basic term as a protective relay, exists in the standards. In various textbooks on relay protection different renditions of this term are provided by various authors, which are far from consistently accurate and which only reflect the subjective views of their authors [7.14]. The lack of a standard definition of the term protective relay encourages not only an arbitrary rendition of this term but also, and this is a consequence, the attribution of functions for which the relay was never designed, which far from being a harmless undertaking could have unforeseen consequences [7.12].

Therefore, in our opinion, standardization in the field of DPR should start with a precise and understandable definition of protective relay, which should undoubtedly be written into the standard. It is obvious that in the decision making process proposed in [7.14] a clearer situation would emerge from the fog that today is descending on relay protection at the hands of some scientists and it would be purged of the ideas and developments which in themselves are very valuable and interesting, but which do not bear any relation to relay protection itself.

7.1.2 The Fundamental Principles of the Standardization of DPR

Which fundamental principles should be reflected in the future standard? In our opinion these should be:

- A ban on using functions in DPR that are alien to relay protection in accordance with a legitimate definition of a *protective relay*.
- A significant limitation on the number of functions in a single microprocessor terminal: a calculation of the optimal number of these functions according to criteria around not only the cost of the protective relay, but also its reliability.
- A ban on using algorithms that employ a non-deterministic logic, which enable the relay to behave in an unpredictable manner.
- The optimum simplification of the programme interface based on some kind of a universal programme platform for all DPR (the standard should set out the fundamental requirements as well as the principles of this platform).
- The introduction of requirements for the reliable operation of DPR under the influence of intentional destructive electromagnetic threats, both by enhancing the immunity of the DPR themselves to these threats, using technical means, that are capable of dramatically attenuating them, and by introducing a secondary protective relay complex for emergency situations, for which an electromechanical relay would be suitable.

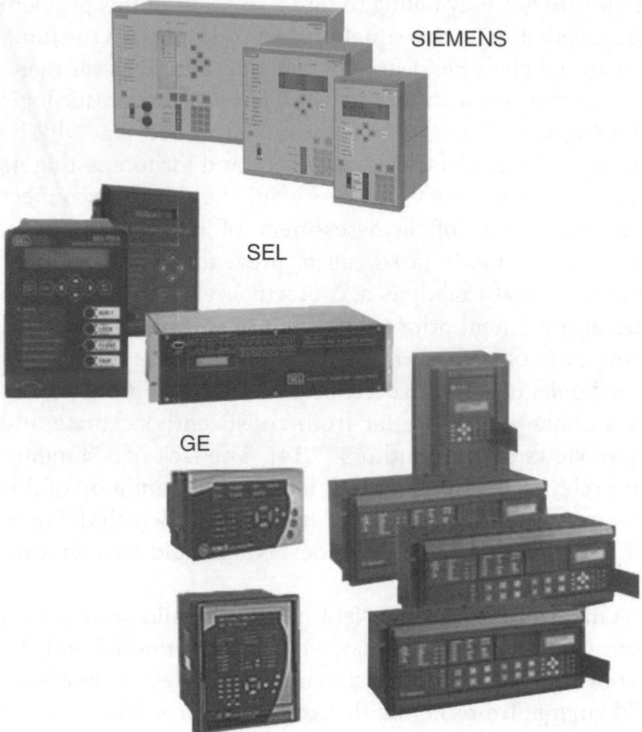

Fig. 7.1 Different types of modern DPR that have different cases and dimensions.

- The introduction of more stringent requirements for cyber security including a ban on the use of technology by which the relay protection commands and signals can be intercepted and intentionally corrupted, such as Wi-Fi technology, and network technology such as Ethernet.

Aside from the general principles set out above in our opinion requirements for the design of DPR should be written into the standard. What kind of requirements does this cover?

7.1.2.1. Standardization of the Design of a DPR

As demonstrated previously, today each type of DPR has its own casing which differs completely from another type of DPR, sometimes from one and the same manufacturer, see Fig. 7.1. These individual DPR are as a rule located today in relay cabinets: 3–5 each in a single cabinet, see Fig. 7.2.

Historically, a situation has come about [7.15] by which today an enormous number of absolutely non-interchangeable designs of DPR that are incompatible with one another are available. Having spent a pretty penny on acquiring a DPR unit from one of the manufacturers, the consumer in actual fact ends up in an economic bondage with this manufacturer over the course of 10–15 years, because once the deal has been concluded with the manufacturer the fact of the variety of manufacturers on the market is of no importance as the consumer is not able to avail themselves of other

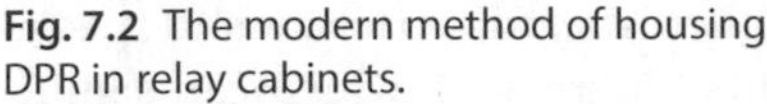
Fig. 7.2 The modern method of housing DPR in relay cabinets.

manufacturers. The only way out of this bondage is to spend no less a sum of money on acquiring a DPR from another manufacturer (and actually to go from one bondage to another).

What would a manufacturer do though in an absolute monopoly? That's correct: They increase the price! The price of a single spare module for a DPR could reach up to a third or even a half of the cost of a far from inexpensive DPR! Since the consumer has nowhere to go they would buy it at this price. What happens then after 8–10 years of exploitation of the DPR? This is what happens: the manufacturer would have assimilated several new designs in that space of time and it becomes unprofitable to maintain production capacity to manufacture spare modules for old relays and the manufacturer simply stops producing them. What is the consumer forced to do in this situation? That's right: throw away the old DPR, even if only one of the modules have failed (the printed boards in modern DPR are manufactured using technology that does not envisage their repair) and fork out money acquiring a new one. Thus the lack of a standard for the design of DPR grows into a serious economic problem that delays the development and modernization of relay protection.

In our opinion, a next generation DPR should be manufactured in the form of separately functional modules (printed boards) unified in terms of dimensions and fitted with unified connectors (couplings). In this case the separate casings become superfluous for printed boards of this nature (in any case in the majority of situations encountered in electrical power engineering). Each DPR can be formed of a horizontal section in a relay cabinet with printed circuit board guides, an individual door, and a rear bulkhead with connectors and terminals to connect external cables. The relay cabinet itself should be manufactured using a special technology, which is designed to protect its contents from electromagnetic threats. Technology exists today (in terms of special cabinets, electrically conductive wiring and lubricants, filters etc.), which can significantly weaken the influence of external electromagnetic emissions in a broad spectrum of frequencies entering highly sensitive apparatus such as a DPR. These cabinets are produced by companies such as R.F. Installations, Inc.; Universal Shielding Corp.; Eldon; Equipto Electronics Corp.; European EMC Products Ltd; Amco Engineering, and many others.

7.1.2.2 Standardization of the Functional Modules in a DPR

Today modules, which make up DPR, are far from always separate functional modules but often take the form of a 'hotchpotch', where different function blocks are located on a single printed circuit board [7.16]. This design is not suitable for realizing the idea of the universalization of DPR, therefore a printed circuit board for a future DPR should take the form of a monofunctional module, such as: A central processing module, a power supply module, an analogue inputs module, a logic inputs module and output relay modules.

If this design solution was used for DPR new 'players' would appear on the market, one of which would specialize in the manufacture of analogue input modules with current and voltage transformers, while others would specialize in the production of the central processor module, a third on digital input modules, and a fourth on the production of different capacity cabinets: from small suspended cabinets to full-scale floor mounted examples. The consumer would be able to put their own DPR together from modules produced by different manufacturers in exactly the same way as personal computers are put together today, taking into account the cost and the quality of these modules. Furthermore, not only would many of the current problems with DPR be solved but the cost of a protective relay would be reduced significantly. The latter would enable two devices of identical protection systems to be fitted in place of one in order to enhance reliability and to enable the second protection system to be used as a spare, which would be activated automatically by a 'watchdog' signal from a damaged main DPR. It would be possible to reject the use of an individual power supply for each DPR and would make it possible for a single dual supply module to be used with an enhanced output and reliability across the cabinet as a whole. It would enable the installation of a number of different service modules in a cabinet such as this, which would enhance the operational reliability of a DPR.

The work of operational personnel, that is to say protective relay servicing personnel, would be simplified dramatically since they would not need to study thick files (Fig. 7.3) for each of the types of DPR installed and define the idiosyncrasies of each of them. Apart from a significant easing of the load on those working on DPR and a reduction in the time needed to assimilate new protection systems, the percentage rate for mistakes caused by the so-called 'human factor' would be significantly reduced.

7.1.2.3 Standardization of the Software used in DPR

The software used in DPR should also be realized, in our opinion, according to principles that have proved themselves well in personal computers, that is to say there should be a base software shell analogous to Windows (but significantly less complex of course) together with an application software as well as libraries designed for specific types of protection functions. If a universal software platform was used together with a unified

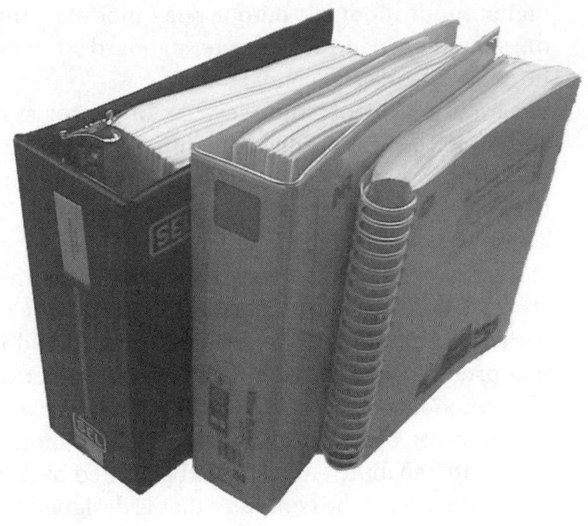

Fig. 7.3 The files running into many pages containing a technical description of a DPR intended for the consumer.

modules design for a DPR a market for application software for different types of protection systems would inevitably appear. Furthermore, it would be possible to achieve the standardization of the interfaces for this application software so that the consumer would not need to convert to a new interface and study it all from scratch every time they purchased a new DPR or new software, as is the case today.

7.1.2.4 On the Need for Standardization in Testing DPR

Testing of a DPR on introduction into service and in operation represents a complex of sufficiently complicated and important operations, and not only does the time expended on the testing depend on the precision and accuracy with which they are conducted but also whether the relay actually works properly. This latter circumstance is supported by the fact that often the settings for a relay need to be changed in the course of testing and conflicting functions need to be blocked and then brought back into operation. A check of the protective characteristics of a complex nature (specifically the characteristics of distance relay protection) for modern DPR is only possible using special modern testing equipment, which itself is very complicated. The need to marry complex DPR correctly with complicated testing equipment represents another issue [7.4], which is examined in detail in [7.14].

7.1.2.5 The Fundamental Principles for the Design of DPR: The Basis of a Future Standard

Thus fundamental principles should in our opinion be written into a future standard under the provisional title of 'The Principles for the Design of DPR: The Fundamental Principles', which are as follows:

1) Functional DPR modules should be separated clearly and the chaotic principle by which these modules are located on printed circuit boards that prevails today [7.15] should be replaced by a more orderly layout stipulated in a special standard. By way of an example: functional modules such as the power supply module, input current and voltage transformer modules containing initial signal processing elements,

digital input modules, output relay modules, the central processor module and so on, should be made on separate, standard sized printed circuit boards that have been fitted with universal connectors.

2) Separate relay protection devices for power systems should not be manufactured and sold as separate individual units, fitted with individual casings of different shapes and sizes, but as individual universal printed circuit boards (modules) from which the consumer would be able to put together a DPR of the required configuration. These boards (modules) should be designed for ease of installation (by placing them along the guide rails up until the point where they join the connector linking to the cross board) and housed in an individual DPR cell in a relay cabinet and equipped with separate doors. The relay cabinets should be manufactured using technology that protects their contents from external electro-magnetic emissions.
3) The functions of DPR should be limited to those of relay protection and nothing else. The number of functions in a single DPR should be optimized in accordance with 'cost' and 'reliability' and should be limited by a standard.
4) The software for the computer that is designed to work with a DPR should consist of a standard base shell and a set of different application programmes and libraries that are compatible with this universal base shell.
5) The power supply to all the modules in the relay cabinet should be conducted from two sources with enhanced reliability, connected together in parallel as a main and a reserve source.

Is it possible from a technical perspective to realize this proposed concept for the manufacture of a DPR? As we have already seen the majority of the DPR available on the market today do not have a set of modules that have been separated strictly according to function, and their design is more reminiscent of a 'hotchpotch', in which the central processor unit is placed on a common printed circuit board along with the switching power supply module, see Fig. 7.4.

However, an analysis we carried out of many different types of modern DPR produced by the leading global manufacturers enabled devices to be found on today's market, which fully meet the requirements set out here in terms of their design [7.16].

Fig. 7.4 A unified DPR module in which the central microprocessor unit is placed together with the switching power supply module and the output relay.

These devices are series 900 DPRs produced by the well-known Chinese company Nari-Relays with their universal modules that are used in different protection systems, see Fig. 7.5. These modules come ready for use and do not require any preparation (aside from configuring the protection programme naturally enough). No debugging of the DPR is required after assembly, which consists solely of installing the printed circuit boards illustrated in Fig. 7.5 (a power supply module is included in the actual set, which is not required in our concept and as such is not shown) in the specially marked guide casing (in our case this would be the cabinet compartment). No more than 10 min are required to assemble such a complex distance protection system made up of seven separate modules supplied in cardboard boxes and to power up the relay, after which the settings can start to be fed into the system. It is completely obvious that an ordinary engineer or relay protection specialist, who does not have any special knowledge in

Fig. 7.5 A set of universal functional modules (22 × 145 mm each) manufactured on different printed circuit boards, and from which different DPRs are put together, those manufactured by Nari-Relays: PCS-931 (differential line protection), a PCS-902 (distance protection), and so on. 1 – the current and voltage input transformer modules; 2 – a narrow band filter (an anti-alias filter); 3 – the logic inputs module; 4 – the output relay module; 5 – the optical communications module and 6 – the central processor module.

the field of microprocessor technology, would cope easily with assembling a protective relay made from these universal modules at the very location in which it is to be installed.

In principle there is nothing to stop the proposed concept being realized within a particular country. On acquiring Nari-Relays' universal modules in the first instance (that use different algorithms written into the EEPROM and which also use different set of input transformers) and after having assimilated construction of the cabinets for these relays even a small company would be able to enter the DRP market today and offer the consumer a new, cheap and reliable protective relay concept, fitted with auxiliary modules.

What are the advantages for this proposed approach to the development of DPR?

For the consumer:

- A significant reduction in the cost of a DPR at the point of purchase;
- The possibility of assembling a DRP from separate modules, from different manufacturers that represent the closest match to the requirements of the consumer from the point of view of the optimum balance between quality and cost;
- The possibility of creating the optimum spare parts for the DPR modules;
- The issue of the reduced reliability of DPR becomes less urgent owing to a fast and convenient replacement of worn out modules on site thanks to the installation of reserve modules that are automatically brought on stream as soon as the principle relays fail; there is no longer any need to repair failed DPR modules;
- The opportunity to opt out of an obligation to a monopolist-manufacturer that sold them the DPR in the first instance;
- Enforcing competition between producers thanks to new market players, small and medium-size companies specializing in the production of only certain types of modules, rather than complete DPR;
- Simplification of DPR testing and a reduction in the influence of the 'human factor';
- It becomes significantly easier to work with the software, and the consumer has the option of selecting the most suitable and convenient application software (interface), which also offers a painless software (interface) transition for one and the same DPR;
- An acceleration of technical progress in the DPR field without making their operation any more complicated and without any additional problems arising for the consumer during a transition to the next generation of devices;
- A reduction in the cost of upgrading DPR since replacing the entire DPR every 10–15 years, which is often the case today, would not prove necessary. It would be sufficient to renew its individual modules. Furthermore, the central processor unit (CPU) could be replaced more frequently than is the case today, thereby accelerating the technical progress in this field.

For the manufacturer:

- There is no need to manufacture obsolete modules, that are required to support the operation of old DPR models;
- No obligation to offer free repair over the entire life cycle of the DPR;
- A significant growth in demand for individual modules;
- The advent of a new application software market (interface shell);
- The option to specialize in the production of some module types, such as the most lucrative for any given manufacturer;

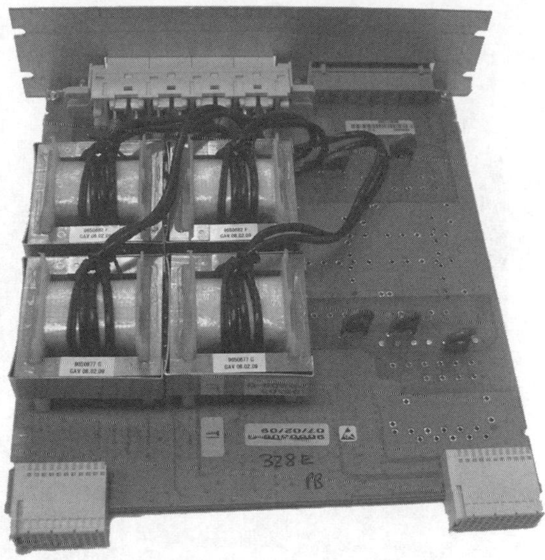

Fig. 7.6 A traditional design of analogue input module with current transformers.

- The option for small and medium size companies to participate in this business that do not have the resources to develop and manufacture complete DPRs;
- A competitive edge for national manufacturers who were the first to commence production of modular DPRs within the territory of a given country over any foreign companies.

The concept of manufacturing a DPR from a set of universal modules does not hinder their upgrading or development in any way or the use of new design principles.

For example, the replacement of what today are standard current transformers in an analogue input module (Fig. 7.6) with miniature shunts (Fig. 7.7) is, in our opinion, very far-sighted. These miniature shunts are manufactured as a bridge with a length of 5 mm between the current inputs located directly on the power connector. This entire connector is divided into four groups with three pins in each. The central pin in each group is common to the current and voltage circuits. The far left hand pin is a current input, while the far right hand pin is a voltage input. That is to say each of the four measuring channels can be used either as an analogue current input, or as an analogue voltage input.

Each of these inputs should be additionally configured as a current or a voltage input with the help of the switchable bridges on the printed circuit board. Despite the miniature dimensions these shunts demonstrate excellent parameters:

- A nominal current of 5 A
- A long time overload of 10 A
- A short time overload (2 s) of 100 A
- An input current range of 0.25–100 A
- An accuracy of 0.1 off range
- A resistance of 3MΩ and a load factor on the external CT – less than 0.1.

In this module the analogue input signals (both current and voltage signals) are transformed into millivolts, which are then transformed into a high-frequency (up to 20 MHz) digital signal by way of AD7401A type analogue-digital convertor with

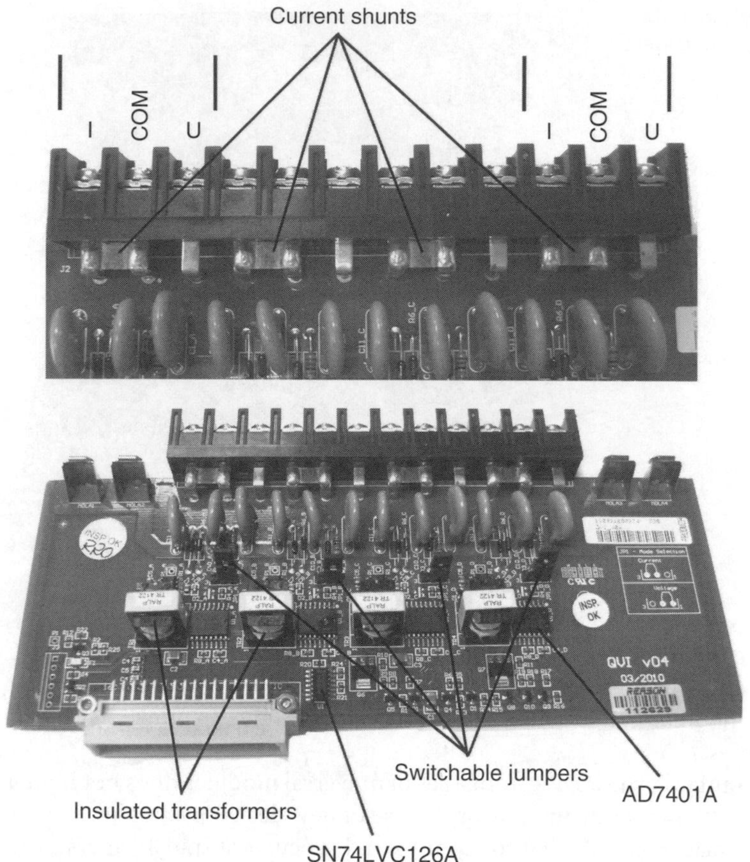

Fig 7.7 An analogue input module in a Digital Recorder RPV-311 (RT Measurement Technologies GbmH) with dimensions of 190 × 75 mm.

high-voltage input-output insulation (according to the manufacturer this is able to withstand a 5 kV voltage for 1 min), and is then passed via the high-frequency transformers with ferrite cores and is normalized with the help of a quadruple SN74LVC126A type logic element and enters the output connector.

Thus the acceptable solutions used by RT Measurements Technologies GbmH have made it possible to create a universal (current/voltage) analogue signals module, that is characterized by its simplicity and its small dimensions, that is to say to take one step closer to the concept discussed above of a set of universal modules to produce different types of protective relays.

It is completely obvious that all of the above is just an outline of some of the general principles that would form a future standard. A wide circle of specialists should be drafted in for the real work on this standard, representing scientists and the future DPR manufacturers as well as future consumers, and representatives of the design organizations. The lack today of such standards, that is to say of limiting frameworks and of any direction for the development of DPRs is today leading to significant economic losses, and in the near future could lead to real chaos in this field.

7.2 The Fundamental Principles for the Standardization of DPR Testing

The standardization of DPR testing is another of the most important directions in reducing the time taken to restore relay protection following the replacement of a DPR that has been hit by an electromagnetic pulse, or by other remote electromagnetic impacts. After the fact of HPEM damage to the power system (substation) and of its emergency shutdown has been established the serviceability and operational condition of the DPRs that are to work again should be tested. One of the problems of DPRs lies in the complexity of testing their serviceability, which inevitably takes its toll on the time it takes to restore the electricity supply.

It is accepted that the serviceability of relay protection devices should be tested using the same fixed settings that are to be used subsequently during the actual operation of the relay at the actual point in the network. As the settings are changed in the process of operation of the relay the serviceability of the relay with these new settings should be checked again. In the days of electromechanical protective relays this was a fully tried and tested measure since a transfer from one setting to another was carried out by way of a mechanical displacement in the relay's internal elements or a switching of the in-built transformer-taps and so on. When the settings in these relays were being changed it was entirely possible that the relay's internal circuits that had been switched to a new transformer-tap, would fail (due to a broken wire, the loss of a contact, or damaged insulation and so on), or it was the case that the new position that the internal elements had been switched to would put it out of balance and it would 'mash' or some other headache would occur. Therefore, the normal serviceability of an electromechanical relay working on one set of settings would not guarantee its serviceability under other settings.

In DPRs the transfer from one set of settings to another is not accompanied by physical changes in its internal structure. Independently from specific settings and operating modes the same input and output circuits operate inside a DPR, the same logic elements, the same processor, the very same power supply and so on. Even the activation and deactivation of the individual functions of the DPR are not connected to any changes in the physical condition of its circuits. A check of the justification for the choice of protection logic and in the justification for the design settings for specific conditions or for a specific circuit – this is a completely different task, which does not bear any relation to checking the serviceability of relays and is not solved by personnel operating the relay and whom are responsible for its serviceability, but by an engineering service responsible for the design settings and the choice of the internal logic by which the relay operates. It is not even possible during the process of testing the serviceability of the relay to model all the situations that might occur in reality or the entire combination of factors active in a real network. It is not the aim of testing the serviceability of protective relays to diagnose these situations. Furthermore, it is possible to demonstrate that a refusal to check a relay using the design settings is a positive measure that reduces the risk of the relay failing as a result of the so-called 'human factor' (which is the reason for almost 50% of the failures in relay protection). The point being that in the multifunctional, digital protection systems the settings for specific conditions are selected in such a way that checking specific functions is only possible by desensitizing or completely disconnecting another, competing function. Not returning this desensitized or disconnected function to its default setting after completion of testing is often the reason for failures in the protection system in emergency modes. An analogous approach to the issue of testing protective relays is proposed in [7.17]. In this document, which in terms of its status is treated as a standard, all the testing of the relay is divided into types: calibration testing (which is designed to test the settings and configuration of the relay) and functional testing. If the periodicity of *once in 4 years for all types of relays* (including electromechanical and microprocessor relays) is set for functional testing, then a periodicity of once in 4 years is set *solely* for electromechanical relays for calibration testing. A periodic calibration (that is to say a checking of the settings) is not envisaged at all for DPRs.

7.2.1 A New Look at the Problem

On the basis of that set out previously, certain principles can be drawn up, which can be adopted for testing DPRs:

1) It is not compulsory to conduct the DPR testing using the settings by which the relay would subsequently operate in a given network in order to confirm its serviceability following repairs or in the process of periodic testing.
2) In order to check the serviceability of DPRs it is sufficient to only check if they are operating correctly in certain very critical areas *that have been predetermined*, at certain very critical (combined) operating modes *that have been predetermined*, including dynamic operating modes with *predetermined* transition processes that are characteristic of typical electrical circuits (although this is not compulsory for this specific network).

These tests should encompass all the relay's physical inputs and outputs. After the testing of the relay has been completed and its serviceability has been confirmed all the test settings should be automatically replaced by a prepared set (file) of actual design settings.

1) Testing of DOR in the most complicated operating modes enables, in our opinion, the most thorough check of the serviceability of the DPR, more so than a limited test within a very limited range of specific settings under which the DPR would subsequently operate.
2) A comprehensive test of a DPR as it enters operation in its most complex operating modes enables the exclusion of additional testing of the serviceability of a DPR during each change in settings, whilst the relay is in operation.

The principles formulated here enable a new approach to the problem of testing DPRs.

One can surmise that the first tools for testing protective relays appeared at almost the same time as the relays themselves. Naturally they were just as primitive as these same protective relays. Initially they were just calibrated inductivity coils, see Fig. 7.8, and rheostats.

As relays developed the systems used to test them became more complex. Test rigs appeared (Fig. 7.9) containing a set of inductances and active resistances, with which it was possible to set the angles between the current and the voltage across a wide range and to check sufficiently complex electromechanical relays.

Different timescales for the periodical checking of protective relays have been set up in different power systems (once in 2–3 years) but these have of course not been strictly observed.

Fig. 7.8 A set of inductances produced by General Electric for checking DPRs.

Fig. 7.9 A TURH-20 type (ASEA) testing system for testing electromechanical relays containing a set of inductances and active resistances.

With the advent of DPRs on the market the situation underwent a cardinal change. The manufacturers of these devices announced that the DPR would allegedly not require periodic checking since it has a powerful in-built self-diagnosis system. This idiosyncrasy of DPRs was featured in advertising brochures as almost their principle advantage over electromechanical and analogue electronic relays. A powerful advertising company hired by the DRP manufacturers played their part. Many specialists in the relay protection field believed this advertising gimmick wholeheartedly and were not able to check the validity of this assertion in practice, although it was completely obvious that it would be impossible to create a testing system based on an internal DPR microprocessor that would check the physical serviceability of many thousands of electronic components. The serviceability of for example an input or output module could not be checked functionally either, without energizing these modules and checking the reaction of the relay to the signals being fed into it. In practice it seems that the majority of DPRs simply do not notice that an entire printed circuit board of one type has been changed for that of another, which is incompatible with that relay's current settings. Many of the author's previous publications concerning this topic have mentioned this and other advertising 'tricks' linked to 'self-diagnosis' in DPRs.

In contrast to the DPR manufacturers the manufacturers of testing systems have always made the assertion that all protective relays should undergo periodic testing including DPRs, since no more than 15% of the software and hardware is covered by the so-called 'self-diagnosis' in these relays. Despite the assertions made by the DPR manufacturers concerning the inexpediency of periodic checking of protection systems, testing systems manufacturers continued to intensively develop more and more modern testing systems and launch them on the market unabated.

7.2.2 Modern Testing Systems to Test Protective Relays

Since the principles of manufacturing DPRs today have become common to the majority of manufacturing companies it is natural that the testing systems available on today's market and which are manufactured by different companies would be very similar to each other, and not just externally, see Fig. 7.10, but also in terms of their characteristics. Today, protective relay testing systems are fully computerized devices that do not have

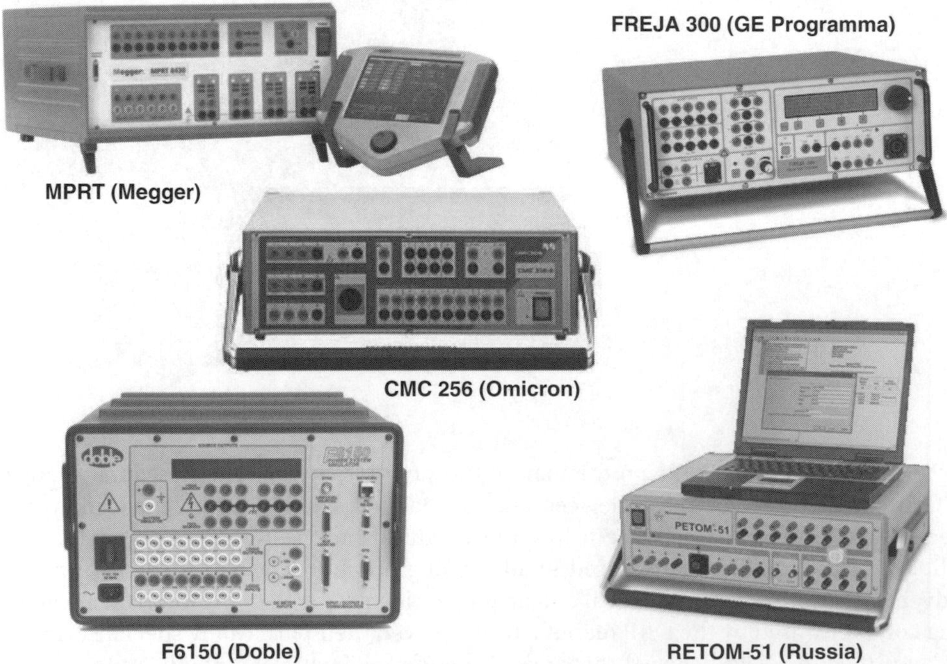

Fig. 7.10 Modern computerized testing systems for testing multifunctional microprocessor based relays.

controls on the front panel, apart from sockets for attaching external cables and an RS232 connector to connect a computer. The cost of these protective relay testing systems runs into tens of thousands of dollars.

These systems are designed to carry out tests across three groups: steady state tests, dynamic tests and transient tests. The first group of tests supposes a test of the relay's base activation settings and represents a preliminary test of the relay in a sense. The second group of tests is designed first and foremost to check how complex protection systems are behaving such as the distance and differential protection in various response regions and protection zones as the input parameters are changed (such as the current, voltage and the angle). The third group of tests supposes an injection of transient process files in the COMTRADE format into the relay's input circuits, that have been drawn from a recording device that records the actual transient process for a short circuit within the network, or of files in this same format that have been manufactured synthetically using special programmes. The test results are fed into a database, which as a rule is based on Sybase SQL Anywhere and are automatically formulated into a standard protocol that can be sent to a printer. The manufacturers of protective relay testing systems normally offer a set of testing procedures (libraries) in the form of macros for the different types of testing and even for some of the widely used types of DPRs.

7.2.3 The Problems with Modern Protective Relay Testing Systems

Modern protective relay testing systems possess an inherent super-flexibility as well as the widest functional capabilities. These protective relay testing systems enable almost any working conditions that protective relays encounter in practice to be simulated, including the synthetic COMTRADE files created in accordance with their own requirements; an artificial distortion of the curved current waveform; a harmonics simulation; a displacement of the current sine wave about an axis (a simulation of the aperiodic component); a simulation of the circuit breaker's initial response; automatic formation of the most complex polygonal characteristics of distance protection systems; a synchronization of differential protection systems via satellites and so on. Super capabilities of this nature found in modern protective relay testing systems gives rise to both sides of the same coin: the need to input hundreds of parameters into dozens of tables in order to carry out each individual test of a relay. Furthermore, the libraries built into the test procedures do not help much either since they do not remove the need to fill a great number of tables. It is worth adding to this the considerable flexibility and universality of the test object (a DPR), which also requires an enormous number of parameters from dozens of drop down menus and tables to be input into it. The smallest inconsistency in the settings for the DPR or relay testing systems leads to inaccurate results. Furthermore, it is not possible every time to ensure that the results obtained are not false by any means. Even in cases when the mistake is obvious (e.g. when the characteristics that have been obtained for a relay do not correspond to the theory) it is very difficult to define exactly where the mistake was allowed to happen: in the DPR settings or in testing system settings. Based on his own experience the author can confirm that searching for a mistake of this nature is very complicated and requires a great deal of effort and time. Working with the Power System Model, which is employed in certain types of protective relay testing systems to check distance protection systems is no less complicated. A knowledge of many of the parameters of an actual electrical network is required to set the parameters for a testing system, which have to be entered into numerous tables with special coefficients. It is often the case that technicians and even relay protection engineers are not aware of these parameters in an actual network or their applicable coefficients and this means that engineers from other energy system services have to participate in the relay checking procedure.

7.2.4 A Proposed Solution to the Problem

It has been long established by psychologists that the greater the numbers of buttons and levers (real or virtual, that is to say software based) that the operator needs to manipulate the lower the effectiveness of a person's interaction with the given technology. Many of the functions and capabilities of this 'fancy' technology are just lost to human understanding. How can the universality and the broadest functional capabilities of a protective relay testing system be compatible with an average technician's capabilities or those of a relay protection engineer who needs to check a limited number of relay types quickly and accurately? Indeed, a testing system manufacturer, overcoming enormous difficulties must develop and fine tune their own procedures and to create their own library of macros based on these procedures make provision for this? We have a series of proposals for a more radical solution to this problem:

1) Modern digital protective relay testing systems of the last generation are technically inexpedient and have not proved themselves economically in terms of their use in testing the simplest of electromechanical relays, such as current voltage relays (like the RT-40, RN-54, IAC type relays as stipulated by the Russian manufacturer of the RETOM-51 type testing system). With this in mind it is significantly more effective to use simpler testing systems. There is no sense in developing test procedures for computerized, automated testing of these types of relays if there is no mention at all of testing hundreds of identical relays during the manufacturing process.
2) The use of in-built libraries of testing procedures in modern digital protective relay testing systems, which require an enormous number of parameters and the knowledge of a multitude of coefficients, can only be considered expedient for old style complex electromechanical protection systems (such as LZ-31 distance protection for example).
3) In order to test modern complex multifunctional DPRs a software platform that is common to all testing systems should be developed, the requirements for which should be inscribed into an international standard. Sybase SQL Anywhere serves as an example of just such a common software platform, and is commonly used to develop databases in various manufacturing and data processing devices, as well as simulators, and testing equipment produced by a number of manufacturers. Another example is the universal COMTRADE format, which is used in all types of digital emergency mode recorders, and specifically in all types of testing systems for simulating transient modes.
4) Different types of applied software for working with protective relay testing systems can have completely different interfaces, but they should all be manufactured in accordance with a common standard software platform.
5) DPR manufacturers should supply their protection in two compact discs. All the settings either for specific operating modes for the protection system should be recorded onto one of these discs under the corresponding numbers, or for the

characteristic response points, or typical examples of electrical networks. The full set of corresponding settings for testing systems as well as diagrams of external DPR connections to the testing system's inputs and outputs must be recorded onto a second compact disc.

6) The effective use of modern testing systems for testing modern multifunctional DPRs is only possible, in our opinion, if the entire testing procedure leads to the settings under number XXI being loaded into the DPR, or the settings under number YYI being loaded into the testing system, and the DPR is connected to the protective relay testing system and...it makes the coffee.

Thus, the proposed set of measures to unify the software platform for modern DPR testing systems of the last generation enables the task of testing modern, multifunctional DPRs to be administered in a completely new way. This in our opinion removes a mass of technical and psychological barriers and would support a significant reduction in the time taken to check the serviceability of DPRs, furthermore, this can be done both under the influence of HPEM and in the standard operating modes for protective relays.

7.3 Establishment of Reserves of Electronic Equipment Replacement Modules as a Way to Improve the Survivability of the Power System

7.3.1 Optimizing the Capacity of Reserves of Replacement Modules

One of the more effective ways to enhance the survivability of an energy system is to quickly restore damaged devices using spare parts, tools and accessories (SPTA). However, building up reserves of SPTA requires considerable funds especially when it comes to the most complex electronic DPRs, automation and control systems that are used widely in power systems. Therefore, a global search has been ongoing for some considerable time for the optimum reserves of SPTA that would enable a blend of the required reliability for these systems with the minimum cost.

Building up the optimum reserves of SPTA is a common problem, that is well-known across many technical fields, which on a theoretical level has been well developed using different mathematical optimization methods [7.18–7.25]. Well-known methods for optimizing reserves of SPTA are based on an analysis of failure statistics for elements, plug-in modules or complete units. That is to say the number of SPTA kits required is calculated based on the fact that failures in electronic equipment are individual random events that occur with a specific frequency that is subordinate to the statistical laws of random values. The need to increase the number of SPTA kits with the aim of ensuring the restoration of serviceability in a certain piece of equipment after the effects of HEMP is not in any doubt. But how can this be increased? It is completely obvious that when it comes to the impact of HEMP on an energy system this results in mass failures in electronic equipment that do not confirm to any statistical laws. Apart from that ordinarily the sufficiently long drawn out process for ordering and receiving new SPTA kits to augment existing reserves once all the pre-prepared kits have been used is not appropriate in this case. Therefore, the simple augmentation of warehouse reserves of SPTA by up to one and a half times or double the reserve (as is sometimes practised by some not very far-sighted managers) does not solve the problem, and an augmentation of reserves by which the entire electronic apparatus would be warranted in terms of SPTA whilst in operation is not viable either from an economic perspective. Therefore, a completely new approach should be used for calculating the optimum SPTA kit.

The proposed method is based on three fundamental principles:

1) Not all the electronic devices should be equipped with SPTA kits, but only those that are defined as critically important devices (CID), without which even the partial function of a power system object is not possible, and from which the critically important sites (CIS) within power systems should be selected.
2) Full and not partial SPTA kits should be made up for the designated CID.
3) Reserves of SPTA kits for CID should be made up in addition to and outside of any connection to the SPTA reserves stored in warehouses.

Thus the optimization of reserves of SPTA in this case comes down in the end to a calculation of the required number of CID necessary to equip the CIS in an energy system.

7.3.2 The Problem of Storing SPTA Reserves

The problem of storing reserves of SPTA requires two problems to be solved: where to store the SPTA and how to store it. Today, in many power systems SPTA kits are stored in warehouses, which are often located far from power system sites and from where maintenance personnel, or personnel directly engaged in operations, would obtain them as required. When a CID requires repair after the impact of HEMP on the energy system the problem of the urgent delivery of critically important cargoes to critically important sites that have been subject to the impact of HEMP as there is every likelihood that the microcontrollers that control modern transport would have been taken out of action.

For telecommunications systems a two tier SPTA system is used in accordance with [7.19], which incorporates the 'SPTA-0' and 'SPTA-G' kits. The SPTA-0 kits form an integral part of the device (in the case of the critically important devices) and should be housed at the same location where the device is being used (in this case at the critically important sites). The SPTA-G kits are stored in a large technical service centre (or at a warehouse). The SPTA-0 and SPTA-G kits should be checked and tested before being handed over for storage.

This approach to organizing the storage of SPTA for electronic apparatus for when the power system needs to be restored after the impact of HPEM is fully justified and is, in our opinion, the most effective approach even in electrical power engineering since it removes the problem of transporting SPTA to a critically important site to restore a critically important site in an emergency. Storage of an SPTA kit for a CID at critically important sites means that this problem does not need to be solved. Another problem could be solved using this approach to storing SPTA for CID: the problem of their configuration and set up prior to their installation into equipment, which would require additional time, as well as the participation of highly qualified personnel using special electronic equipment and computers (which could also be damaged). Modern DPRs could serve as an example of a critically important device of this nature, without which power systems are not able to function. Given a large-scale failure as a result of the impact of HPEM it would be very difficult to set up a large number of DPR SPTAs

Table 7.1 The minimal requirements for effective shielding of critically important assets from the impact of HEMP (based on Fig. 1 in the MIL-STD-188-15-1 ([6.26]).

Frequency of emissions	Attenuation inserted by the shield, dB
10 kHz	20
100 kHz	40
1 MHz	60
10 MHz	80
1 GHz	80

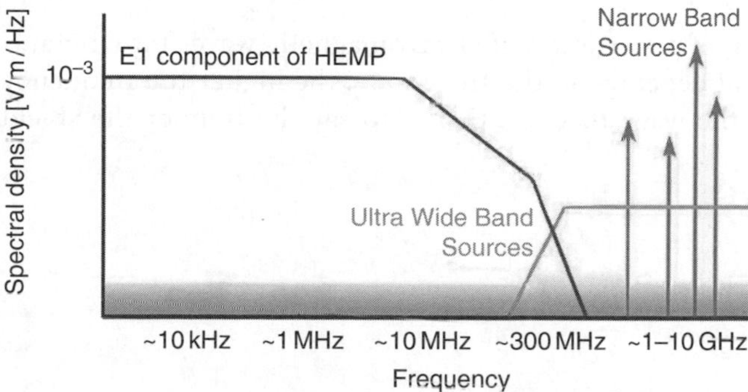

Fig. 7.11 The spectral density of emissions for different sources of HPEM (from the standard IEC 61000-2-13).

simultaneously across a number of remote power system sites. Therefore, the SPTA kits for DPRs that have been designated CID should not only be stored near to active DPRs but should be pre-programmed, tuned and configured to enable a swift replacement of the failed modules in a given DPR (the plug-and-play principle), which operates under these same settings.

The second issue that needs to be solved is: how should SPTA for a critically important device be stored? The problem is that HEMP creates an electrical field strength at a level that can reach up to 50 kV/m. Given this field strength a voltage differential could arise between pins in even relatively small electronic components (placed in a single PCB module) sufficient for an electrical breakdown in the p-n junctions, in the thinnest layers of insulation in microprocessors or the wiping of information in the nucleus of the memory cells. Therefore, the SPTA kits for a critically important device should be stored in containers protected from HEMP.

What properties should these containers possess? Our attention is drawn to the MIL-STD-188-125-1 [7.26], which specifies the requirements for effective shielding of critically important assets from the impact of HEMP, see Table 7.1.

At the same time the IEC-61000-2-13 [7.27] standard sets out data concerning the spectral density of the emissions from different types of HPEM (see Fig. 7.11) from which it is evident that for the E1 and E2 components (in the standard E2 is specified as

Table 7.2 The depth of penetration of an electromagnetic wave into a shield made from aluminium across different frequencies.

Frequency	1 kHz	10 kHz	100 kHz	1 MHz	10 MHz	100 MHz	1 GHz
Depth of penetration of the wave, mm	2.6	0.83	0.26	0.083	0.026	0.0083	0.0026

'lightning' since its parameters are closest to those of lightning) of HEMP the density of the emissions remains at its highest at frequencies of lower than 10 kHz and drops sharply at frequencies above 300 MHz. Other sources of HPEM (as opposed to HEMP) create a relatively high emission frequency across a higher frequency range, right up to 10 GHz. Therefore, a frequency range from units of kilohertz to 10 GHz should provide effective protection.

As is well known, the depth of penetration of electromagnetic waves into metal is defined by the skin-effect and depends on the frequency: the higher the frequency (f) the lower the depth (Δ) the wave reaches, that is to say the thinner the shield can be:

$$\Delta = 503\sqrt{\frac{\rho}{\mu_m f}}$$

Where:

Δ - the depth of penetration of the wave
ρ - the specific resistance of the metal
μ_m - the magnetic permeability of the metal
f - the frequency of the emissions

The depth of penetration of an electromagnetic wave is defined as the thickness of the surface layer of the metal, in which the electromagnetic field strength is reduced to e = 2.718 times. According to data [7.28] almost 86% of the energy that passed through the surface would escape in this layer. In Table 7.2 the results of a calculation using the formula given here are set out for a metal that is most commonly used as a shield; aluminium.

As is evident from the table the aluminium container with a thickness for the shield of not less than 3 mm is able to provide effective weakening for HPEM emissions of any kind.

What does today's market for protective containers offer? First and foremost, this market is well represented by large metal containers with thick walls, see Fig. 7.12, equipped with protective ventilation and filtration systems for input cables.

These containers are widely used in the army and do unfailingly provide the excellent protection stipulated here for the equipment housed inside them. Unfortunately, they represent very expensive protection and it is doubtful that they could be used to store SPTA for power systems purposes. Another variant of a protective container is a room without windows but with doors and walls, and lined with copper plate (rooms such as

Fig. 7.12 Large metal containers designed to provide protection from HEMP and equipped with ventilation and filtration systems to connect external cables.

this are manufactured by a company called Holland Shielding Systems). Protective containers of this nature also possess excellent shielding capabilities (from 40 to 120 dB in a frequency range of 10 kHz – 10 GHz), but as was the case with the containers above they are very expensive.

Plastic 'Faraday bags' with a metal layer are, according to their manufacturer's assertions, a reliable and very simple type of container that provides protection against HEMP, see Fig. 7.13.

As a rule, the manufacturers of these bags indicate a high degree of weakening of emissions, reaching up to 40–45 dB, but this being the case they are shamefacedly quiet about the frequency range for which this data has been obtained. Can a metallized layer with a thickness of several microns effectively attenuate an electromagnetic field with a frequency range from kilohertz to gigahertz? Table 7.2 provides a definitive answer to this question: no, it cannot!

Another type of protective container that is also widely represented on the market and which is advertised as a reliable protection device against HEMP is a tent, which is manufactured from the same metallized plastic as the bags, or in the best case scenario is woven from material containing metallic threads, see Fig. 7.14.

Special transient thick walled metallic containers that provide very effective shielding are also well represented on the market, see Fig. 7.15.

Unfortunately, these containers are too expensive to store SPTA and their internal capacity is not sufficient. In our opinion the most suitable are simple, aluminium containers welded from sheet aluminium in the form of simple boxes with lids, see Fig. 7.16. These containers with a thickness in the walls of 3/16ths of an inch (4.8 mm) demonstrate an entirely acceptable level of shielding: not less than 50 dB in a frequency range

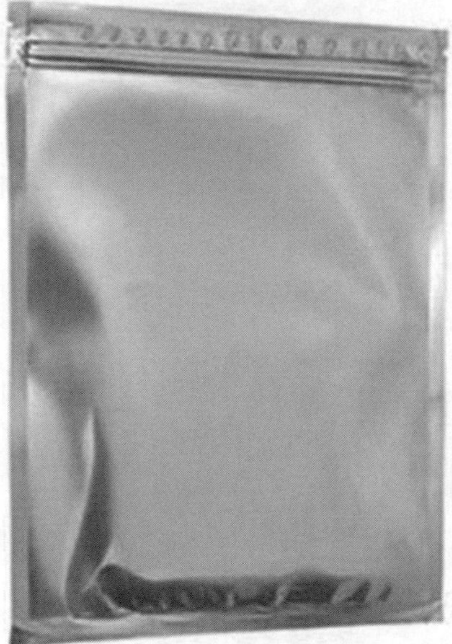

Fig 7.13 Plastic 'Faraday bags' of different sizes with a metallized layer that were designed to protect small electronic devices from EMP.

Fig. 7.14 A protective tent made from metallized cloth.

of 100 kHz – 1 GHz (76 dB at a frequency of 300 MHz; 66 dB at a frequency of 1 GHz) and are manufactured by a series of companies including Montie Gear, EMP Engineering, and others in standard or bespoke sizes.

It is worth noting that these simple containers can be manufactured to the required size in any workshop that undertakes welding work. Furthermore with the aim of countering the impact of electromagnetic fields at the upper reaches of the frequency range on the electronic devices held inside (which could penetrate into the container's internal cavities via gaps formed in the loose fitting lids) it is recommended that especially

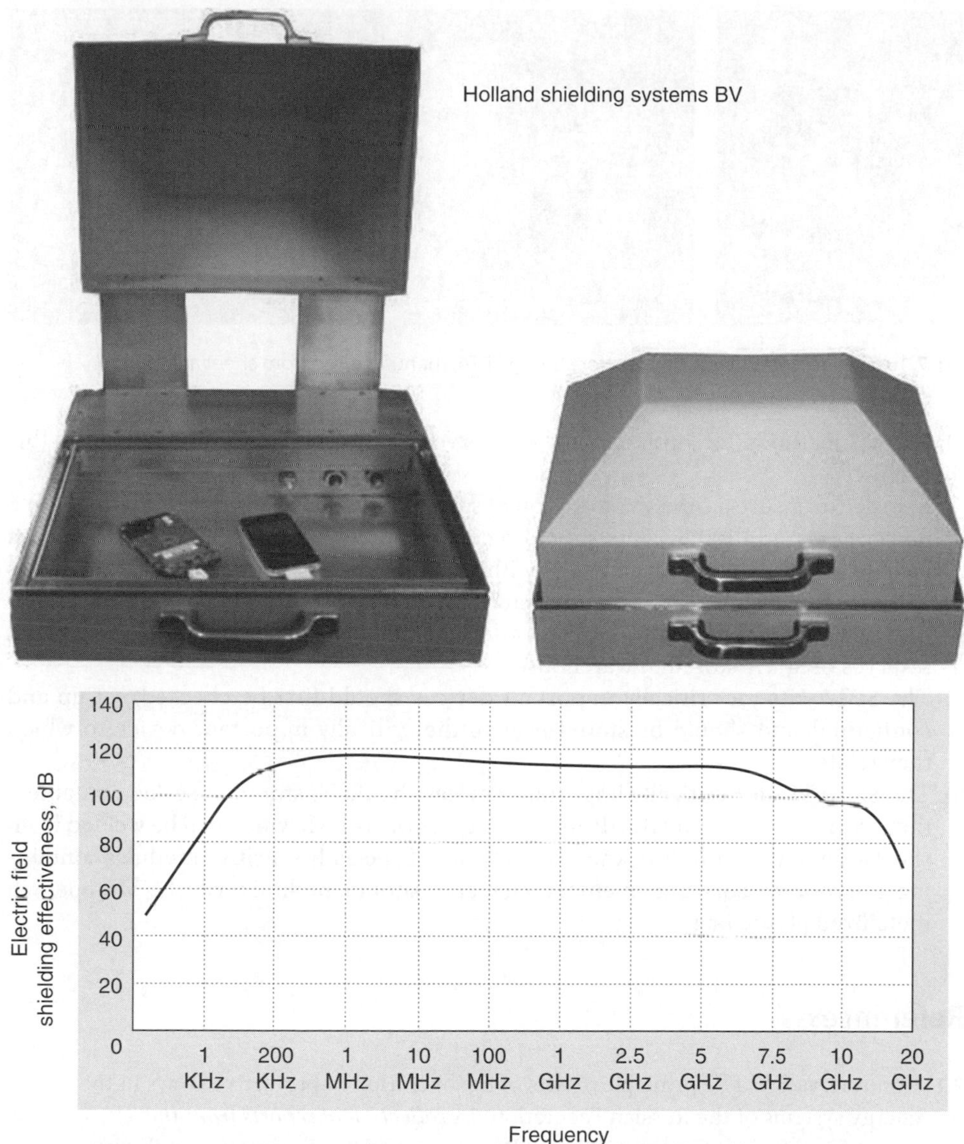

Fig. 7.15 A protective container produced by Holland Shielding Systems BV that possesses a very high shielding capability.

sensitive electronic equipment (such as printed circuit boards with microprocessors and memory elements) should be wrapped in the metallized plastic bags described previously before they are placed in containers.

Summing up, the following conclusions can be made:

1) One of the measures for a timely restoration of serviceability for an electrical power system after the impact of one or another of the types of HPEM is to create special SPTA kits for electronic equipment.

Fig. 7.16 Cheap protective containers for storing SPTA, manufactured from sheet aluminium.

2) Known methods for optimizing the reserves of SPTA are not acceptable in this instance.
3) In order to ensure a timely restoration of serviceability for electronic equipment in a power system, full SPTA kits should be developed for critically important devices located at critically important sites within power systems, both the critical devices and the critical sites should be pre-determined.
4) The SPTA kits for critically important devices should be independent of the general reserves of SPTA stored in warehouses.
5) The SPTA kits for critically important devices should first be checked, set up and configured, and should be stored near to the critically important device to which they relate.
6) The SPTA kits for critically important devices should be stored in sealed containers that are protected from HEMP and other types of HPEM, which can be welded from sheet aluminium with a thickness of just 5 mm. Especially sensitive modules containing microprocessors and memory elements should preliminarily be wrapped in metallized plastic bags.

References

7.1 Konovalova E.V. The principle results of the operation of protective relays in the energy systems of the Russian Federation - *A collection of reports from the XV Scientific-Technical Conference 'Relay Protection and the Automation of Energy Systems'.* Moscow, 2002.

7.2 Kjolle G.H., Heggset J., Hjartsjo B.T., Engen H., Protection system faults 1999–2003 and the influence on the reliability of supply - *2005 IEEE St. Petersburg Power Tech, St. Petersburg, Russia*, June 27–30, 2005.

7.3 Gurveich V.I. Problems in assessing the reliability of relay protection. - *Electricity*, 2011, No. 2, pp. 28–31.

7.4 Gurevich V.I. Testing DPRs - Electro: electrical technology - *Electrical Power Engineering. The Electrical Power Industry*, 2009, No. 1 pp. 31–33.

7.5 Bittencourt A., de Carvalho M.R., Rolim J.G., Adaptive strategies in power systems protection using artificial intelligence techniques - *The 15th International Conference on Intelligent System Applications to Power Systems, Curitiba, Brazil*, November 8–12, 2009.

7.6 Laughton M.A. Artificial intelligence techniques in power systems. In *Artificial Intelligence Techniques in Power Systems*, IET, 1997, pp. 1–18.

7.7 Khosla R., Dillon T., Neuro-expert system applications in power systems. In *Artificial Intelligence Techniques in Power Systems*, IET, 1997, pp. 238–258.

7.8 Lyamets Y.Y, Kerzhayev D.V., Nudelman G.S., Romanov Y.V., Multifaceted relay protection - Articles from reports presented at the *Second International Scientific-Technical Conference 'Contemporary Trends in the Development of Relay Protection Systems and the Automation of Energy Systems' Moscow*, 7–10 September 2009.

7.9 Kamel T.S., Khassan M.A., EL-Mordshedi A. (Cairo University Egypt) Using artificial intelligence systems in the remote protection of overhead power lines - Articles from reports presented at the *Second International Scientific-Technical Conference 'Contemporary Trends in the Development of Relay Protection Systems and the Automation of Energy Systems' Moscow*, 7–10 September 2009.

7.10 Bulychev A. Nudelman, G. Relay protection. Upgrades as a result of preventative functions - *The Electro-Technical News*, 2009, No. 4(58).

7.11 Gurevich V.I. Fata Morgana or visionaries at the All-Russian Scientific-Research Institute for Flow Metering - *Electrical Networks and Systems*, 2009, No. 6, pp. 54–57.

7.12 Gurevich V.I. The 'Intellectualisation' of relay protection; Good intentions or the road to Hell? -*Electrical Networks and Systems*, 2010, No. 5, pp. 63–67.

7.13 Belyaev A.V., Shirokov V.V., Yemelyantsev A.Yu. Digital relay protection and automation terminals. The practice of their adaptation to Russian specifications - *Electro-Technical News*, 2009, No. 5.

7.14 Gurevich V.I. *The Problems of Standardisation in the DPR Field - SPB*: The DEAN Publishing House, 2015. 168 p.

7.15 Gurevich V.I. Current issues in relay protection: An alternative view - *The Electrical Power Engineering News*, 2010, No. 3, pp. 30–43.

7.16 Gurevich V.I. A new concept in the manufacture of DPRs - *Components and Technology*, 2010, No. 6, pp. 12–15.

7.17 PJM Relay Testing and Maintenance Practices. PJM Interconnection. Relay Subcommittee. Rev. 2/26/04, 2004.

7.18 Zhadnov V. Automation in the design of spare components in SPTA: methods and techniques - *Components and Technology*, 2010, No. 5, pp. 173–176.

7.19 OST 45.66-96 Spare parts, tools and accessories for telecommunication devices. Industry Standard. M.: *The Central Scientific Technical Institute 'Informsvyaz'*, 1997.

7.20 Zatsarinniy A.A., Gagarin A.I., Kozlov S.V. et al. The idiosyncrasies of designing SPTA components in automated information systems in a protected environment - *Systems and Information Technology Devices*, 2013, Vol. 23, No. 1, pp. 113–131.

7.21 Dopira R.V., Lysyuk A.P., Tsybenko D.V. et al. Methods for designing a system to provide spare parts for radio-electronic technology distributed on a territorial basis - *Software Products and Systems*, 2009, No. 1, pp. 128–130.

7.22 GOST RV 20.39.303-98. A comprehensive general purpose system. military apparatus, instruments, devices, and equipment. reliability requirements - *The Composition and Arrangement of the Specification.* - M.: The Standartov Publishing House, 1998.

7.23 Trimp M.E., Dekker R., Teunter R.H. Optimize initial spare parts inventories: an analysis and improvement of an electronic decision tool - *Report Econometric Institute EI 200-52*, Erasmus University Rotterdam, 2004, 70 p.

7.24 MIL-STD-1388-2 Department of Defense Requirements for a Logistic Support Analysis Record, 1993.

7.25 Love R.E., Stebbins B.F. An analysis of spare parts forecasting methods utilised in the United States Marine Corps. - Thesis AD-A184 698, Naval Postgraduate School, Department of the Navy, 1987.

7.26 MIL-STD-188-125-1 High-altitude electromagnetic pulse (HEMP) protection for ground-based C4I facilities performing critical, Time-Urgent Missions; Part 1: Fixed Facilities.

7.27 IEC 61000-2-13: 2005 Electromagnetic Compatibility (EMC) - Part 2-13: Environment - High-power electromagnetic (HPEM) environments - Radiated and conducted.

7.28 Nekrasova N.M., Katsevich L.S., Yevtyukova I.P. *Industrial Electro-Thermal Installations/Gosenergoizdat*, 1961, 415 p.

8

Protecting High-Power Electrical Equipment from EMP

8.1 The Magneto-Hydrodynamic Effect of HEMP

The magneto-hydrodynamic effect of EMP (MHD-EMP) is one of the components of HEMP, which is designated the E3 component. At its heart lies the magneto-hydrodynamic effect of the interaction between the plasma produced by the nuclear explosion and pre-heated, ionized air with the Earth's magnetic field. There is a distinction between the two stages of this effect, which are referred to in foreign literature as the 'Blast Wave' and the 'Heave', both of which have different formation and duration mechanisms, see Fig. 8.1. The first stage with a duration of up to 1–10 s, is brought about by the release of large amounts of plasma material that is formed during the explosion in rarefied air (at high altitude) and in the presence of the Earth's magnetic field. Furthermore, a complex interaction takes place between the plasma ions, and the magnetic field, as well as gamma and X-ray radiation, which is accompanied by the formation of a vortex electrical field.

These physical effects lead to a significant displacement of the Earth's magnetic field, which grows stronger as the energy of the explosion and its altitude above the Earth increases. The heave occurs at the second stage together with the shock-heated and heavily ionized air caused by the explosion, that is to say plasma in actual fact, moving upwards rapidly. As the ionized plasma dissects the field lines in the Earth's magnetic field it is accompanied by a polarization of the air layer and the generation of a powerful electrical field, which in turn forms powerful circulating currents in the ionosphere. These processes are relatively slow. The duration of this phase of the explosion is 10–300 s.

The result of all these processes in the shock-heated atmosphere is the advent of a slowly varying electrical field at the Earth's surface with a field strength ranging from units to dozens of volts across a kilometre, see Fig. 8.2.

Despite the small field strength in the electrical field brought about by the E3 HEMP component, it induces in long-distance metallic objects (pipes, rails, OPLs) relatively powerful electrical currents with a very low frequency (less than 1 Hz), that is to say quasi-direct currents [8.3]. The currents that are induced in OPLs are especially dangerous, see Fig. 8.3.

Protection of Substation Critical Equipment Against Intentional Electromagnetic Threats,
First Edition. Vladimir Gurevich.

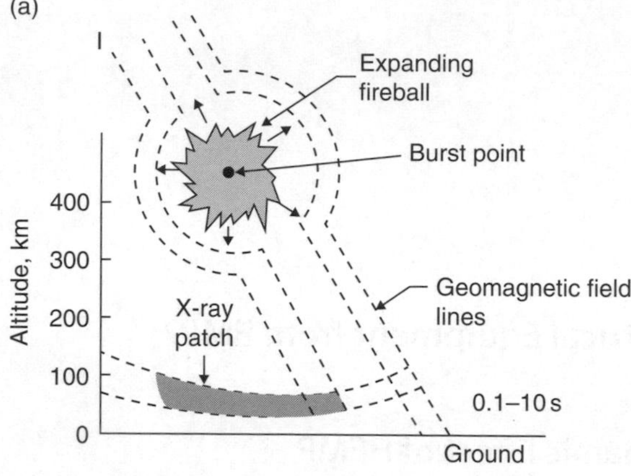

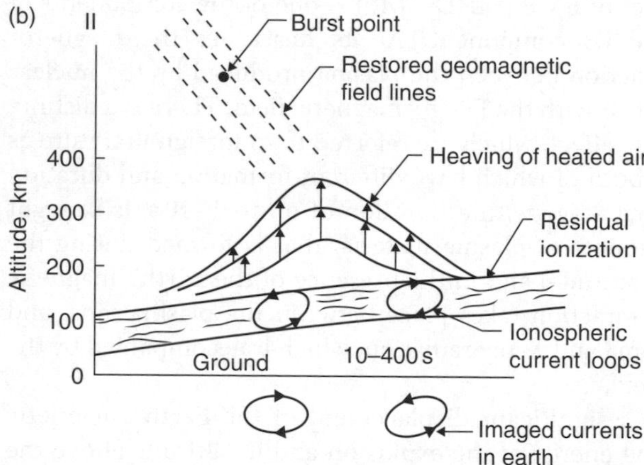

Fig. 8.1 The two stages of the magneto-hydrodynamic effect of HEMP [8.1]: (a) the 'blast wave' and (b) the 'heave'.

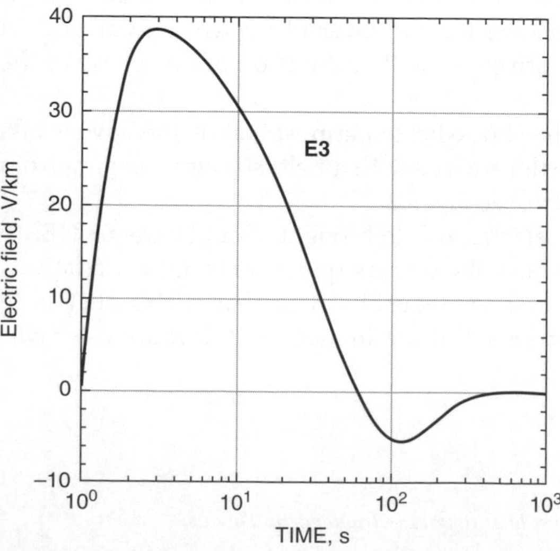

Fig. 8.2 The change in the field strength of the horizontal component of the electrical field on the Earth's surface as a result of the impact of the E3 HEMP component [8.2].

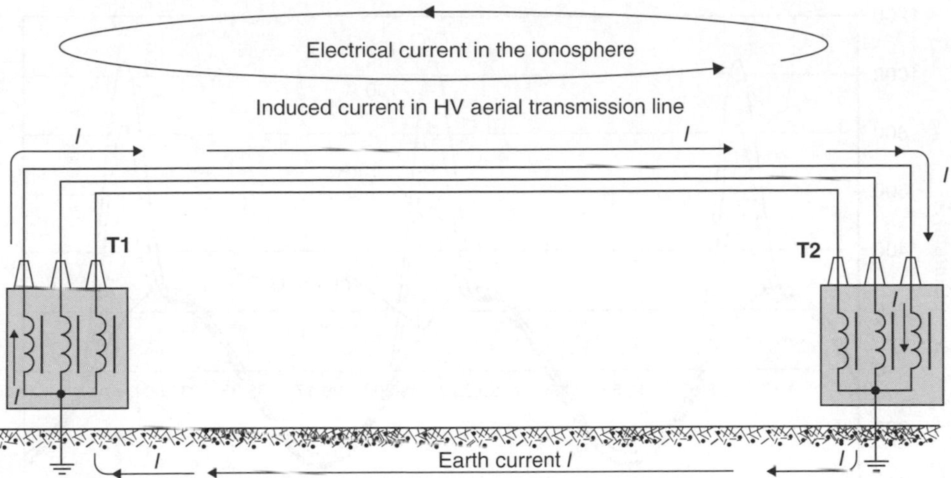

Fig. 8.3 A diagram showing the circulation of currents, induced in the wires of OPL and which are looped via the neutrals in high-power transformers.

8.2 The Influence of the E3 HEMP Component on High-Power Electrical Equipment

Since the geo-magnetically induced current (GIC) has a very low frequency (less than 1 Hz) its impact is analogous to that of direct current and would manifest itself first and foremost in electrical equipment containing electromagnetic systems such as high-power transformers. The saturation of the core of a high-power transformer by a quasi-direct current in the neutral leads to an increase in the magnetizing current, a significant current curve distortion in the transformer's windings, and to a significant increase in losses in a transformer and an increase in the temperature of the winding and of the magnetic core, see Fig. 8.4. For a transformer the danger of this operating mode lies not only in the likelihood of a failure, which is high, but also in its negative impact on the entire power system.

A transformer that is operating in this mode is a powerful source of even and uneven harmonics, causing an overload of the capacitive compensation battery and destroying the normal operation of a protective relay. The protection devices for the bank of capacitors switch them off, protecting them from the overload. Combined with a simultaneous dramatic increase in consumption of reactive power by the transformer itself, that is operating in this mode, and given its considerably high power, this leads to a significant deficit in the reactive power in the system as a whole. Furthermore, the voltage is reduced and the voltage regulators (transformer tap changers) that are attempting to restore the voltage level automatically come into operation. The on-load tap changer contactors (OLTC) are devices that are not designed for switching currents containing a significant DC component and there is every likelihood that they would be destroyed, which could lead to them sustaining damage and to a short circuit in the transformer windings. The circuit breakers should switch off momentarily in this operating mode, de-energizing the damaged transformers. Here is the real question though, would the high-voltage circuit breakers be able to disconnect these short circuit currents and would they in any way be able to de-energize load currents that contain a high DC component?

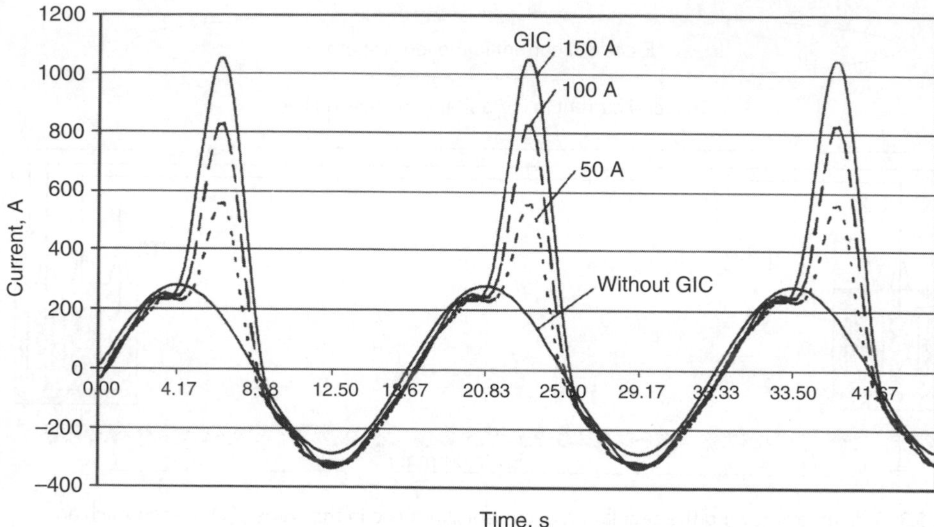

Fig. 8.4 The distortion of the shape of a steady-state current in transformer windings on the influence of geomagnetic currents with an amperage of 50 100 and 150 A as they penetrate a neutral circuit.

They are not designed to switch DC currents of this nature. What happens then to the capacitors shunting the poles for these circuit breakers that are connected in series when they are under the influence of high-frequency harmonics? There are still more questions than answers. Nevertheless, it is well-known that powerful solar storms, which cause effects that are similar in their parameters to the effects caused by the E3 HEMP component have led more than once to high-power electrical equipment becoming seriously damaged and to the collapse of the energy systems in a number of countries.

8.3 Protection of High-Power Equipment from the Impact of Geo-Magnetically Induced Currents (GIC)

It is completely obvious that it is possible to protect the established high-power electrical equipment found in energy systems effectively, by preventing GIC from passing through them. This can be achieved either by preventing GIC from entering the system via OPLs (by using a series capacitance battery inserted in the wires of OPL), or by blocking GIC from entering the neutral inputs in high-power transformers (by inserting capacitors into the neutral earth circuit in series).

A series capacitance battery for OPL is an expensive luxury and is only used in very long distance OPLs to compensate for the inductivity of the wires. Therefore, recent times have seen the intensive development of apparatus based on a capacitor designed to prevent GIC from penetrating the neutral points of high-power transformers.

The problem is that in the absence of GIC this apparatus should not have any influence on a transformer's normal operating modes or that of the electrical network, that is to say they should not reduce the effectiveness of the neutral earth, and should be able to withstand large short circuit currents, and furthermore should prevent the triggering

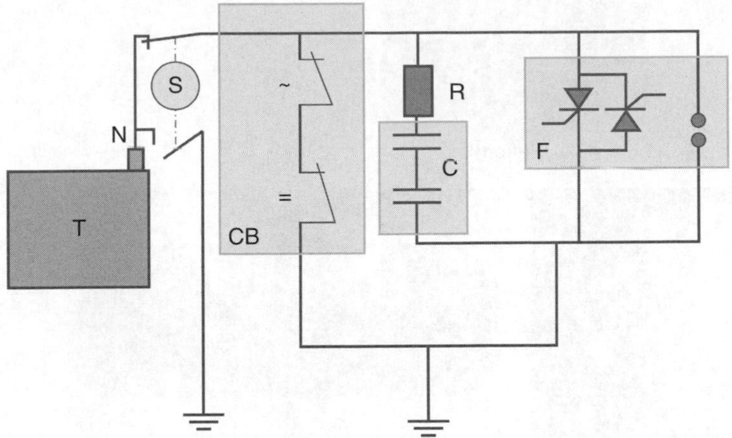

Fig. 8.5 A typical diagram of a device that blocks the GIC in the neutrals in high-power transformers. T – the transformer, S – switching apparatus, designed to make the device inoperative, CB – a separate circuit breaker that is designed to break alternating and direct currents, C – bank of capacitors, R – limiting resistor, F – special device that provides protection from overvoltages in cases where high-level emergency currents enter the neutral circuits.

of ferro-resonant phenomena and overvoltages in transient modes. With the aim of enabling this an operating algorithm has been accepted for use in all apparatus of this nature in which the bank of capacitors is constantly being shunted by a bypass switching element (CB) and works normally by de-shunting (disconnecting this switching element) at the moment the GIC is detected, see Fig. 8.5.

The advent of emergency modes with capacitors built into the transformer neutral could lead to the advent of very high voltages in the neutral circuits within transformers that exceed the insulation level for the neutral, as well as on the capacitors themselves. Therefore, apparatus of this nature should be equipped with special devices to protect them from the overvoltages F (standard varistors are not suitable for this task). The protection device consisting of modules formed of six high-power high-voltage thyristors and a vacuum spark-gap is illustrated in Fig. 8.5. In practice high-power triggered triple electrode vacuum spark-gaps of a special design are often used in place of thyristor modules, see Fig. 8.6.

Apart from that the possibility of de-energizing this equipment using specific switching apparatus (S), such as a disconnector with an earthing knife to turn OFF the device without de-energizing the transformer should be proposed. On the whole this apparatus is not simple or cheap (more than $300 000), see Fig. 8.7.

There are other methods of protecting high-power transformers from GIC, which are based on changes in the design of a transformer. The insertion of additional non-magnetic gaps in the transformer's core reduces the probability of its saturation, but reduces the important technical characteristics of the transformer.

Other well-known technical solutions (Fig. 8.8) envisage the accommodation of additional windings in a transformer's magnetic circuits that compensate for the influence of direct current, shunted by some sort of specific element (31) that has high resistance to DC and low resistance to AC.

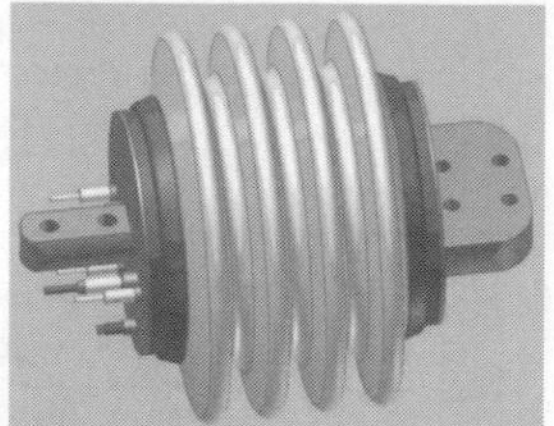

4275 Bi-tron™

3275 Pulsatron™

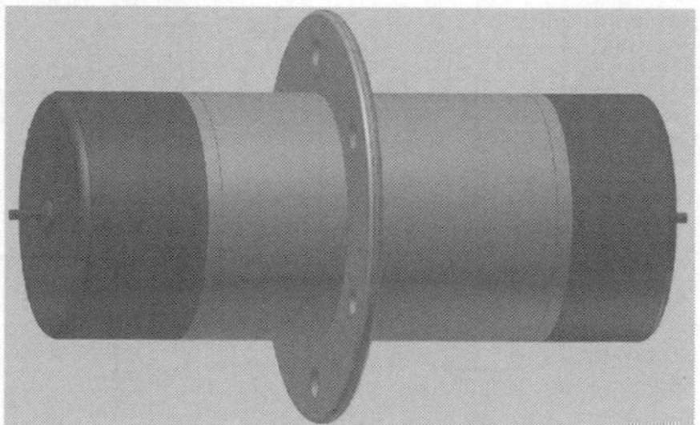

4138 Bi-tron™

Fig. 8.6 High-power high-voltage triggered spark-gaps produced by Advanced Fusion Systems, types: 4275 (35 kV, 100 kA); 3275 (500 kV, 250 kA) and 4138 (75 kV, 250 kA).

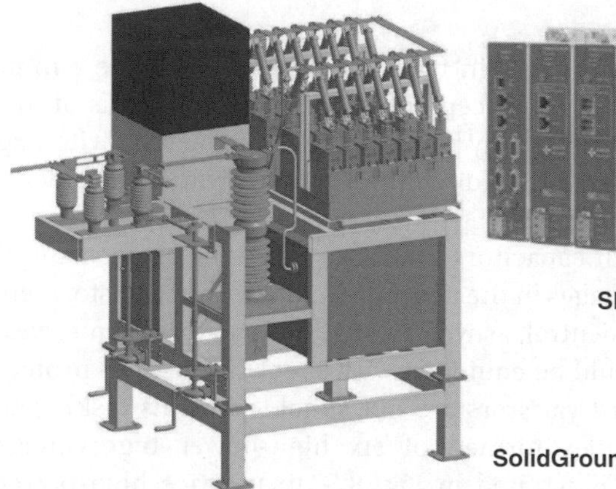

SEL-2240 Axion™

SolidGround™ (ABB)

NBBD (Phoenix Electric Co.)

Fig. 8.7 The equipment designed to block GIC in the neutral circuits of transformers: the top is produced by ABB and is equipped with a controller produced by SEL, below it is a device developed by Phoenix Electric.

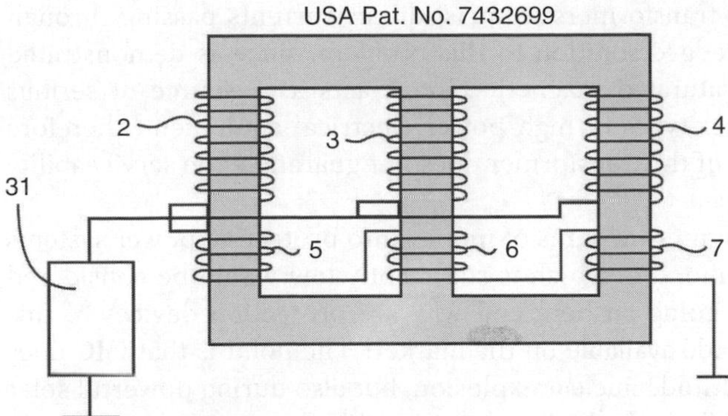

Fig. 8.8 A high-power transformer device with an additional winding, compensating for the GIC, 2–4 – the base windings; 5–7 – Compensatory windings; 31 – an element with a high resistance to DC and a low resistance to AC.

Evidently, a bank of capacitors could serve as this element although this is not mentioned directly in US patent No. 7432699. An analogous technical solution is described in US patent No. 7489485. In other technical solutions it is proposed that this compensatory winding be fed from an external, controlled DC source that compensates the GIC.

There are proposals to connect the transformer's windings in a 'reverse zigzag' pattern, in which the compensation of the magnetizing currents is mutual in each phase, with an equal value, but in an opposite direction, and the magnetic circuits are not saturated.

All these technical solutions require changes in the technology used to manufacture high-power transformers, reduce their technical characteristics, and lead to them becoming significantly more expensive. That is to say any technical measures that are aimed at preventing/compensating for GIC currents, are linked to significant material expenses. In connection with which a question arises concerning the economic expediency of investing significant funds with the aim of preventing damage to electrical equipment during such an exceptional event as a high altitude nuclear explosion.

A series of leading global manufacturers of high-power, high-voltage transformers (such as Siemens, ABB and others) advise that they are developing special transformers (GIC Safe Power Transformers), that are capable of withstanding GIC with a value of up to 50 A over the course of several hours, as well as separate GIC pulses with an amplitude of up to 200 A over the course of several hours. The manufacturers do not give away the technical solutions by which they have been able to enhance the immunity of transformers to GIC in their publicity material. It is obvious however that this is not about technical solutions of some kind that block the quasi-direct currents from entering the neutral points of a transformer and compensating for magnetic flux in a magnetic circuit, since even if these solutions were used there are no significant restrictions on the length of time an exposure to GIC may last, neither are there any specific restrictions on its magnitude.

First and foremost, these transformers simply have an enhanced immunity to a higher temperature thanks to the use of special high-temperature lacquers to insulate the core plates, as well as special insulating materials for the windings.

Enhancing the immunity of transformers to quasi-direct currents passing through them offers a far from fully-fledged solution to this problem, since as demonstrated above a transformer with a saturated magnetic circuit acts as a source of serious problems for many of the other types of high-power electrical equipment, therefore safeguarding the serviceability of the transformer does not guarantee the serviceability of the power system in any way.

It is unlikely that investing significant sums of money into protecting power systems from a one off event, the likelihood of which is sufficiently low would be considered expedient. Therefore, what is going on here and why are protection devices of this nature being developed and made available on the market? The point is that GIC does not only occur during a high altitude nuclear explosion, but also during powerful solar storms, which occur periodically and which are responsible for serious incidents in energy systems. However, the influence of solar storms is not the same in different parts of the Earth. This influence increases closer to the poles and is very weak in regions close to the Equator. In areas located at a significant distance from the poles solar storms do not give rise to any perceptible GIC at all that would be capable of affecting the serviceability of energy systems. However, those who develop protective devices point to circumstances in which the GIC parameters that are caused by solar storms and those caused by a high altitude nuclear explosion are similar to one another in many ways, and as such power systems that require enhancement of the immunity in their electrical equipment to a high altitude nuclear explosion should employ these protection devices independent of their geographical location. It would appear that this is entirely logical.

There is however one important difference in the parameters of GIC caused by a powerful solar storm and one that is the result of a high altitude nuclear explosion, which puts the basis for this logic in doubt. This difference lies in the life span of the GIC. In a high altitude nuclear explosion, the life span of the GIC is limited to a few minutes, during the course of which high-power transformers that possess a high heat capacity are simply not capable of heating up to a dangerous temperature due to a short time period, see Fig. 8.9.

It is understandable that given high volumes of GIC the temperature of parts of the transformer would be higher but in such a short time it would nevertheless not reach figures that would be dangerous for a transformer. Currents that have been distorted asymmetrically by a saturated transformer present a very serious danger for the other kinds of high-power equipment examined before that do not possess the type of inertia found in high-power transformers. However, if during a solar storm the GIC is present over the course of several hours, during which the electrical power supply to consumers needs to be maintained, then during a high altitude nuclear explosion the life span of GIC is restricted to several minutes, during the course of which electrical equipment may be shut down to prevent it from being damaged and it can then be brought back into operation. Furthermore, since the process by which the GIC is built up is sufficiently slow the transformers can be switched off the moment a direct current component is detected in the current in the neutral conductor, without waiting for the magnetic circuits to be saturated and for whatever would follow this saturation. This method for protecting high-power electrical equipment from the magneto-hydrodynamic effect of HEMP seems to us to be significantly better than those examined above, since even though it is highly effective it does not require significant material expense. All the outlay for realizing this protection method boils down to the installation of a special

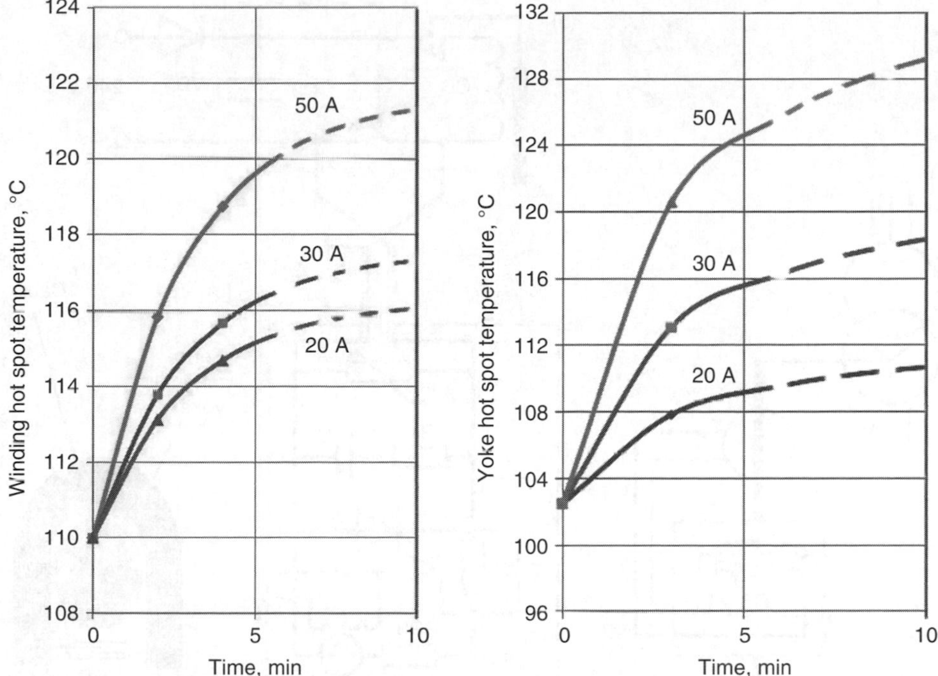

Fig. 8.9 Examples of a power transformer winding (left) and yoke (right) as they are heated under geomagnetic currents of 20, 30 and 50 A flowing in the neutral.

relay that reacts to the advent of a direct current component in the current flowing through the neutral conductor, and which sends an immediate signal to de-energize the transformer.

A relay such as this is of a specific design and is not the same as the one that ABB used to control the SolidGround device (a SEL-2240 type industrial programmable controller) since this refers to a protection device designed to protect from the E3 HEMP component, which appears after the E1 and E2 components and would in all probability, take the SEL-2240 digital control device out of action before it has had a chance to work.

Figure 8.10(a) illustrates the operating principle for a relay that is sensitive to a direct current component in the neutral conductor in a high-power transformer and is insensitive to an alternating current varying across a broad range. This relay consists of an RS reed switch with a winding that is located on a cable connecting the transformer's neutral conductor with an earthing point that is perpendicular to the cable axis and the toroidal current transformer installed on this cable. In the absence of a direct current component in the current flowing through the neutral conductor the cable's magnetic field that acts directly on the reed switch is balanced completely by the magnetic field from the coil that is applied to the reed switch, and which is powered by a current transformer. A change in the alternating current flowing through the neutral conductor leads to a proportional change in both of these magnetic fields that act upon the reed switch and to their mutual compensation. In the advent of a significant direct current component in the current flowing through the neutral conductor (of higher than 10–15 A) the balance in the magnetic fields acting on the reed switch is disturbed: the magnetic

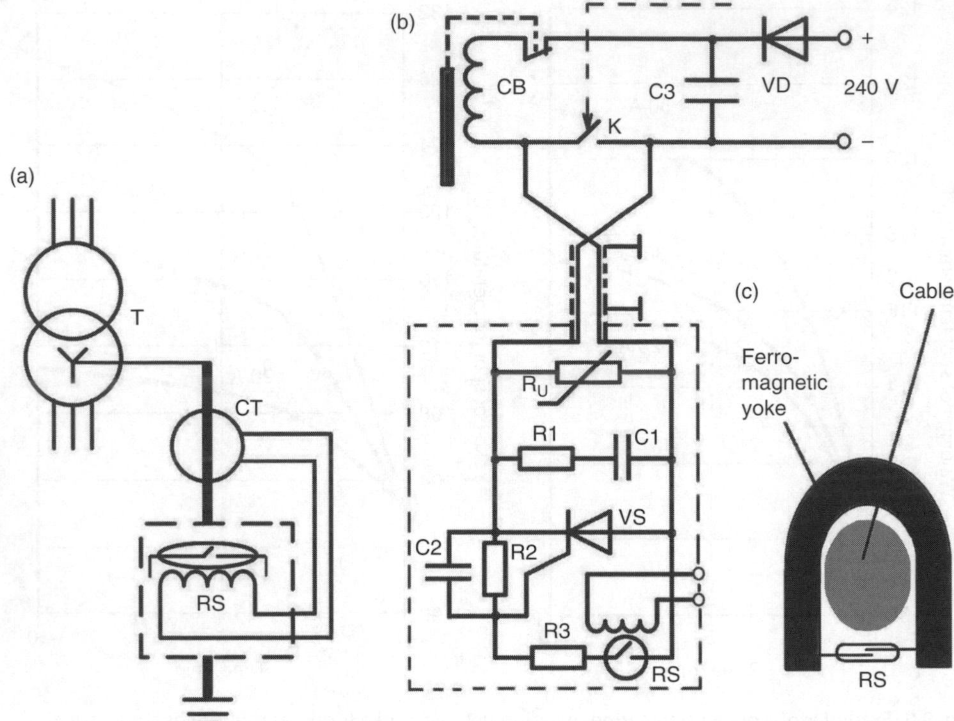

Fig. 8.10 A protective relay for a high-power transformer designed to protect from low frequency induced geomagnetic currents in the neutral circuit.

field in the cable acts on it as before but the compensatory magnetic field from the coil that is energized by the current transformer is absent, since the direct current component is not flowing via the current transformer. As a result, the reed switch is activated. The actual design for the relay incorporates an additional power amplifier on the VS thyristor, the RU varistor and the R1C1 chain that protects the thyristor from the interference and overvoltages, see Fig. 8.10. The relay is equipped with a bulk electrostatic shield and a ferromagnetic shield, and the only window is on the cable side where the reed switch is located and is connected to the circuit by a circuit breaker trip coil CB by way of a special shielded cable with twisted pair cables and a multilayered combined shield, earthed from both ends and which is immune to the impact of EMP. A miniature high-voltage vacuum reed switch of the KSK-1A85 type can also be used in this relay (these are produced by Meder Electronics) and have an insulating strength between the contacts of 4000 V with a bulb diameter of 2.75 mm and a length of 21 mm. This reed switch is capable of switching power of up to 100 W (the maximum switchable voltage is 1000 V; the maximum switchable current is 1 A); the activation time is 1 ms and the maximum sensitivity is 20 A/turns. When necessary additional ferromagnetic cores (magnetic field concentrators) can be used to enhance the sensitivity, which would be accommodated close to the reed switch. In order that the relay possess a higher sensitivity and a higher activation threshold the longitudinal axis of the reed switch should form an angle distinct from 90° to the axis of the cable on which it is mounted. The thyristor has also been chosen as it is a miniature high-voltage SKT50/18E type (produced by Semicron) with a maximum voltage of 1800 V and with a maximum sustained current of 75 A, and is able to withstand a high voltage rise rate (1000 V/mcs) and a high operating temperature range (−40 + 130 °C). The trip coil feed circuit is equipped with a storage capacitor C3, which ensures activation of the circuit breaker even if a control voltage should penetrate it. The RC2C chain is designed to enhance the immunity of the device to interference still further. The C2 capacitor ensures a certain delay in activation of the thyristor preventing it from turning ON under the influence of high-power pulse interference.

The use of discrete high-voltage components in place of the traditional microelectronics in this relay enables it to provide excellent reliability under the influence of the impact of HEMP.

On the basis of this, the following conclusions can be drawn:

1) The magneto-hydrodynamic effect of HEMP (the MHD-EMP) that consists of a quasi-direct current flowing through the neutral conductors of high-power transformers not only has a negative effect on the transformers themselves but also on many other kinds of high-power electrical equipment, first and foremost on the bank of capacitors and high-voltage circuit breakers. Therefore, the technical solutions aimed at protecting power equipment from the influence of MHD-EMP should envisage not only protection of the transformers themselves, but of all the other kinds of high-power electrical equipment in power systems. Technical solutions not aimed at blocking or compensating for GIC but solely at enhancing the immunity of transformers to these currents passing through them cannot be considered effective.
2) Existing technical solutions based on preventing the saturation of the magnetic core in transformers can be preliminarily divided into two groups:
   a) first, use of external apparatus, inserted in the neutral conductors in transformers and blocking the flow of quasi-direct currents in neutral circuits.
   b) second, internal design changes to the transformer itself (in its windings and magnetic circuits) that are blocking the saturation of the magnetic circuits from the quasi-direct current's flow through the neutral circuits.
3) Any of the known technical solutions aimed at maintaining the normal functions of the power system and its elements under the influence of GIC are linked to considerable material expense.
4) Despite the resemblance of many of the parameters of GIC that arise as a result of a high altitude nuclear explosion (the E3 component) and as a result of powerful solar storms, there is one important difference between them: the life span of the GIC. This difference dictates the requirement for differing approaches to the protection of electrical equipment from the impact of solar storms and from the E3 HEMP component.
5) The use of known technical solutions for protecting high-power electrical equipment in power systems from the impact of the E3 HEMP component cannot be recognized as being economically motivated. In this case de-energizing a high-power transformer for a short period of time over a few minutes following a signal from a special relay, with the transformer being subsequently brought back into operation automatically is justified.
6) A relay that transmits a signal to de-energize a transformer should be of a specific design, which ensures its serviceability under the influence of all the components of HEMP.

References

8.1 Study to Access the Effects of Magnetohydrodynamic Electromagnetic Pulse on Electric Power Systems - Report ORNL/sub-83/43374/1/v, Oak Ridge National Laboratory, 1985.

8.2 IEC 61000–2-9 Electromagnetic compatibility (EMC) - Part 2: Environment - Section 9: Description of HEMP environment - Radiated Disturbance, 1996.

8.3 The Late-Time (E3) High-Altitude Electromagnetic Pulse (HEMP) and its Impact on the U.S. Power Grid. - Report Meta-R-321, Metatech Corp, 2010.

Appendix

EMP and its Impact on the Power System

List of Reports

I. EMP Theory

1.1 Dolan P.J. Report DTIC ADA955391: Capabilities of Nuclear Weapons. Part 2. Damage Criteria. Change 1. Chapter 7. Electromagnetic Pulse (EMP) Phenomena. NNA EM-1 – Defense Nuclear Agency, 1978.
1.2 Report DTIC ADA059914: Effect of Multiple Scattering on the Compton Recoil Current – Mission Research Corp. for Defense Nuclear Agency, 1978.
1.3 Lee K.S.H. Report AFWL-TR-80–402: EMP Interaction: Principles, Techniques and Reference Data – Air Force Weapons Laboratory, 1981.
1.4 Lee K.S.H. Interaction Note 435: Interaction of High-Altitude Electromagnetic Pulse (HEMP) with Transmission Lines. An Early-Time Consideration – LuTech Inc., 1983.
1.5 Gould K.E. Report DTIC ADB094426: A Guide to Nuclear Weapons Phenomena and Effects Literature – Defense Nuclear Agency, 1984. Unclassified in 1986.
1.6 Tesche F.M. Interaction Note 458: A Study of Overhead Line Responses to High Altitude Electromagnetic Pulse Environments – LuTech Inc., 1986.
1.7 Longmire C.L., Hamilton R.M., Hahn J.M. Report ORNL/Sub/86-18417/1: A Nominal Set of High-Altitude EMP Environments – Oak Ridge National Laboratory, 1987.
1.8 Blanchard J.P., Tesche F.M., McConnell B.W. Report ORNL/Sub/85-27461/1: The Effects of Corona on Current Surges Induced on Conducting Lines by EMP: A Comparison of Experimental Data with Results of Analytic Corona Models – Oak Ridge National Laboratory, 1987.
1.9 Uman M.A. Report DTIC ADA234306: Comparison of the Frequency Spectra of HEMP and Lightning – Defense Nuclear Agency, 1991.

Protection of Substation Critical Equipment Against Intentional Electromagnetic Threats,
First Edition. Vladimir Gurevich.

II. Geomagnetically Induced Currents and its Impact on Power System

2.1 Legro J.R., Abi-Samra N.C., Tesche F.M. Report ORML/Sub-83/43374/1/V3: Study to Assess the Effects of Magnetohydrodynamic Electromagnetic Pulse on Electric Power Systems – Oak Ridge National Laboratory, 1985.
2.2 Barnes P.R., Rizy D.T., McDonell B.W. Report ORNL-6665: Electric Utility Industry Experience with Geomagnetic Disturbances – Oak Ridge National Laboratory, 1991.
2.3 High-Impact, Low-Frequency Event Risk to the North American Bulk Power System – A Jointly-Commissioned Summary Report of the North American Electric Reliability Corp. and the U.S. Department of Energy's November 2009 Workshop – NERC, 2010.
2.4 Kappenman J. Report Meta-R-322: Low-Frequency Protection Concepts for the Electric Power Grid: Geomagnetically Induced Current (GIC) and E3 HEMP Mitigation – Metatech Corp., 2010.
2.5 Geo-Magnetic Disturbances (GMD): Monitoring, Mitigation and Next Steps. A Literature Review and Summary of the April 2011 NERC GMD Workshop. – North American Electric Reliability Corp. (NERC), 2011.
2.6 Popik T.S. Effect of Geomagnetic Disturbances on the Bulk Power System and Electromagnetic Pulse Effect on the U.S. Power Grid – Task Force on National and Homeland Security, 2012.
2.7 Girsis R., Vedante V., Gramm K. Report A2-304: Effects of Geomagnetically Induced Currents on Power Transformers and Power Systems – CIGRE, 2012.
2.8 Ngnegueu T., Marketos F., Devaux F. Report A2-303: Behavior of Transformers Under DC/GIC Excitation: Phenomenon, Impact on Design/Design Evaluation Process and Modeling Aspects in Support of Design – CIGRE, 2012.
2.9 Samuelsson O. Report LUTEDX/(TEIE-7242)/1-21/(2013): Geomagnetic disturbances and their impact on power systems – Division of Industrial Electrical Engineering and Automation Faculty of Engineering, Lund University, 2013.
2.10 Fernández Barroso C.D. Report LUTEDX/(TEIE-5328)/1-062/(2014) GIC Distribution – Division of Industrial Electrical Engineering and Automation Faculty of Engineering, Lund University, 2014.

III. EMP Impact on Power System

3.1 Marable H., Barnes P.R., Nelson D.B. Report ORNL-4958: Power System EMP Protection, Final Report – Oak Ridge National Laboratory, 1975.
3.2 Manweiler R.W. Report ORNL-4919: Effect of Nuclear Electromagnetic Pulse (EMP) on Synchronous Stability of the Electric Power System – Oak Ridge National Laboratory, 1975.
3.3 Report DTIC ADA014489: Electromagnetic Pulse and Civil Preparedness – Defense Civil Preparedness Agency, Washington, 1975.
3.4 Vance E.F. Report ADA009228: Electromagnetic-Pulse Handbook for Electric Power Systems – Stanford Research Institute, 1975.
3.5 Report HCP/T5103-01: Impact Assessment of the 1977 New York City Blackout. – U.S. Department of Energy, Division of Electric Energy Systems, 1978.
3.6 Ericson D.M., Strawe D.F., Sandberg S.J. Report NUREG/CR-3069: Interaction of Electromagnetic Pulse with Commercial Nuclear Power Plant – Sandia National Laboratories, 1983.

3.7 Zaininger H.W. Report ORNL/Sub/82-47905/1: Electromagnetic Pulse (EMP) Interaction with Electrical Power Systems – Oak Ridge National Laboratory, 1984.
3.8 Barnes P.R., Vance E.F., Askins H.W. Report ORNL-6033: Nuclear Electromagnetic Pulse (EMP) and Electric Power Systems – Oak Ridge National Laboratory, 1984.
3.9 Engheta N., Lee K.S.H., Yang F.C. Report ORNL/Sub-84/73986/1: HEMP-Induced Transients in Transmission and Distribution (T and D) Lines – Oak Ridge National Laboratory, 1985.
3.10 Smith D. Report ORNL/Sub/84–89643/1: Study for a Facility to Simulate High Altitude EMP Coupled Through Overhead Transmission Lines – Oak Ridge National Laboratory, 1985.
3.11 Dethlefsen R. Report ORNL/qub/84-89642/2: Design Concepts for a Pulse Power Test Facility to Simulate EMP Surges in Overhead Power Lines – Oak Ridge National Laboratory, 1985.
3.12 Ramrus A. Report ORNL/Sub/84-89642/1: Design Concepts for a Pulse Power Test Facility to Simulate EMP Surges in Overhead Power Lines: Part 1, Fast Pulse – Oak Ridge National Laboratory, 1985.
3.13 Dethlefsen R. Report ORNL/qub/84-89642/2: Design Concepts for a Pulse Power Test Facility to Simulate EMP Surges in Overhead Power Lines: Part 2, Slow Pulses – Oak Ridge National Laboratory, 1986.
3.14 Legro J.R. ORNL/Sub/83-43374/V1: Study to Assess the Effects of Electromagnetic Pulse on Electric Power Systems – Phase I Executive Summary – Oak Ridge National Laboratory, 1985.
3.15 Legro J.R. Report ORNL/Sub/83-43374/V4: Study to Assess the Effects of Nuclear Surface Burst Electromagnetic Pulse on Electric Power Systems: Phase I – Oak Ridge National Laboratory, 1985.
3.16 Legro J.R. Report ORNL/Sub/83-43374/1/VI: Study to Assess the Effects of High-Altitude Electromagnetic Pulse on Electric Power Systems – Phase I Final Report-Oak Ridge National Laboratory, 1986.
3.17 Taylor E.R. Report No. AST 88-2081: HEMP-Type Impulse Transfer Tests on High-Voltage Bushing Current Transformers at the Maxwell Laboratory – Westinghouse AST, 1988.
3.18 Taylor E.R. Report No. AST 88-7072: HEMP-Type Impulse Tests on High-Voltage Potential Transformers at the Maxwell Laboratory – Westinghouse AST, 1988.
3.19 Report DTIC ADA206952: The Effects of High-Altitude Electromagnetic Pulse (HEMP) on Telecommunications Assets – National Communication System, 1988.
3.20 Liu T.K. Report ORNL/Sub/88-02238/1: HEMP Test and Analysis of Selected Recloser – Control Units – Oak Ridge National Laboratory, 1989.
3.21 Report EP 1110-3-2: Engineering and Design - Electromagnetic Pulse (EMP) and Tempest Protection for Facilities – Department of the Army, U.S. Army Corps of Engineers, 1990.
3.22 Kruse V.J., Nickel D.L., Bonk J.J. Report ORNL/Sub/83-43374/2: Impacts of a Nominal Nuclear Electromagnetic Pulse on Electric Power Systems – Oak Ridge National Laboratory, 1991.
3.23 Burrage L.M. Report ORNL/Sub/85-28611/2: Impact of Steep-Front Short-Duration Impulses on Power System Insulation – Oak Ridge National Laboratory, 1991.

3.24 Report ORNL/Sub/83-43374/2: Impact of a Nuclear Electromagnetic Pulse on Electric Power Systems (Phase III, Final Report) – Oak Ridge National Laboratory, 1991.
3.25 Chrzanowski P., Futterman J. Report CD-90-0014: An Assessment of the Electromagnetic Pulse (EMP) Effects on the U.S. Civilian Infrastructure – Unclassified Summary and Recommendations – Lawrence Livermore National Laboratory, 1992.
3.26 Wiggins C.V., Thomas D.E., Salas T.M. Report ORNL/Sub-88-SC863: HEMP-Induced Transients in Electric Power Substations (final report) – Oak Ridge National Laboratory, 1992.
3.27 Wagner C.W., Feego W.E. Report ORNL/Sub-91-SG913/1: Recommended Engineering Practice to Enhance the EMI/EMP Immunity of Electric Power Systems – Oak Ridge National Laboratory, 1992.
3.28 Barnes P.R., McConell B.W., Van Dyke J.W. Report ORNL-6708: Electromagnetic Pulse Research on Electric Power Systems – Program Summary and Recommendations – Oak Ridge National Laboratory, 1993.
3.29 Reddoch T.W., Markel L.C. Report OPNL/Sub/91-SG105/1: HEMP Emergency Planning and Operating Procedures for Electric Power Systems – Oak Ridge National Laboratory, 1993.
3.30 Barnes P.R., McConnell B.W. Report ORNL/TM – 1999/93: Assessment and Testing of Long-Line Interface Devices – Oak Ridge National Laboratory, 2000.
3.31 Technical Manual TM 5–690: Grounding and Bonding in Command, Control, Communications, Computer, Intelligence, Surveillance and Reconnaissance (C41SR) – Headquarters Department of the Army, 2002.
3.32 Wilson C. Report for Congress RL32544: High Altitude Electromagnetic Pulse (HEMP) and High Power Microwave (HPM) Devices: Threat Assessments – Congressional Research Service, 2008.
3.33 Report of the Commission to Assess the Threat to the United States from Electromagnetic Pulse (EMP) Attack – Critical National Infrastructures – 2008.
3.34 Critical Infrastructure Strategic Initiatives Coordinated Action Plan – Technical Committee Report – North American Electric Reliability Corp. (NERC), 2010.
3.35 Savage E., Gilbert J., Radasky W. Report Meta-R-320: The Early-Time (E1) High-Altitude Electromagnetic Pulse (HEMP) and Its Impact on the U.S. Power Grid – Metatech Corp., 2010.
3.36 Gilbert J., Kappenman J., Radasky W., Savage E. Report Meta-R-321: The Late-Time (E3) High-Altitude Electromagnetic Pulse (HEMP) and Its Impact on the U.S. Power Grid – Metatech Corp., 2010.
3.37 Radasky W., Savage E. Report Meta-R-323: Intentional Electromagnetic Interference (IEMI) and Its Impact on the U.S. Power Grid – Metatech Corp., 2010.
3.38 Radasky W., Savage E. Report Meta-R-324: High-Frequency Protection Concepts for the Electric Power Grid - Metatech Corp., 2010.
3.39 Oreskovic R. Strategy Research Project: Electromagnetic Pulse – A Catastrophic Threat to the Homeland – U.S. Army War College, 2011.
3.40 Report HC 1552: Developing Threats to Electronic Infrastructure – House of Commons, Defence Committee, 2011.
3.41 Report HC 1552: Developing Threats Electro-Magnetic Pulses (EMP) – House of Commons, Defence Committee, 2012.

3.42 Report 5200.44: Protection of Mission Critical Functions to Achieve Trusted Systems and Networks (TSN) – U.S. Department of Defense, 2012.
3.43 Critical Infrastructure Protection Committee Strategic Plan – North American Electric Reliability Corp. (NERC), 2012.
3.44 Report 3002000796: Electromagnetic Pulse and Intentional Electromagnetic Interference (EMI) Threats to the Power Grid: Characterization of the Threat, Available Countermeasures, and Opportunities for Technology Research – Electric Power Research Institute (EPRI), 2013.
3.45 Report 113–85: Assured Microelectronics Policy – U.S. Department of Defense, 2014.
3.46 Beck C. E-Pro Report (International Electric Grid Protection) – Electric Infrastructure Security Council, 2013.
3.47 Report FM 3–38: Cyber Electromagnetic Activities. – Headquarters Department of the U.S. Army, 2014.
3.48 Report of CIGRE C4.206 Working Group: Protection of High Power Network Control Electronics Against Intentional Electromagnetic Interference (IEMI) – CIGRE, 2014.
3.49 Report INL/EXT-15–35582: Strategies, Protections and Mitigations for the Electric Grid from Electromagnetic Pulse Effects – Idaho National Laboratory (INL), 2016.

Index

Protection of Substation Critical Equipment Against Intentional Electromagnetic Threats, First Edition. Vladimir Gurevich.